Acknowledgements

I0020294

I wish to thank my colleague Gerard Parr for all his help during the years of my Doctorate research. I am forever indebted to my parents for all their encouragement and their financial support. I'd like to thank my wife for her never-ending patience – "I'm just leaving the office...right now, see you in ten minutes... honestly!". Finally to my two sons Jack and Levi.....may this be but one of many books that you read in your long, peaceful and beautiful lives.

Over the past 30 years, telecommunications has enjoyed three major changes, each a real epoch. The first in the '70s was digital. The second was packet switching. The third was wireless. Each spurred some fundamental innovation. The first launched multimedia, among other things. The second, "always on" connectivity. And the third, functional mobility. Combined with worldwide deregulation, these generational changed have given the consumer more services at lower costs.

Nicholas Negroponte, *Founder of MIT Media Lab, "Being Wireless", Wired Digital October 2002*

Table of Contents

List of Figures

Preface

Seamless connectivity to multiple wireless networks independently of a fixed point is becoming increasingly important for mobile devices, but wireless networks differ in bandwidth, size and access costs each requiring protocol functions to enable devices to communicate efficiently. Portable multimedia devices such as PDA's with restricted memory and smart phones will also vary greatly and all these devices will require optimal multimedia delivery. Traditionally, sources limit transmission rates to accommodate lower bandwidth links, even though high-bandwidth connectivity is available to many participants, but this does not provide optimum throughput to all clients due to its quest for a common denominator bandwidth. In addition, due to the divergence of applications, traditional protocol stacks are frequently enriched with additional functionality such as transport protocol functionality, and presentation coding which can lead to a performance bottleneck due to the insufficient processing power and memory of portable devices.

This book presents an extensible middleware which caters for small resource constrained devices to full-fledged desktop computers hence dynamic micro-protocols are investigated which enable devices to adopt specific protocol stacks at runtime in an attempt to optimise the stack to the functionality that is actually required by the application, thus eliminating additional overhead functionality provided by generic stacks. A side effect of this is that it allows devices such as PDAs to offer protocol functions, which would not normally be available due to their memory constraints. Memory constrained devices are catered for through the deployment of a client-proxy overlay network where proxies offload essential processing. The problem of the 'common denominator bandwidth' is overcome through multicast media groups, where clients subscribe to different quality of services in accordance with resource availability and can move between groups according to bandwidth availability over time. The result is a 100% Java middleware for streaming to heterogeneous mobile clients, utilising dynamic configuration of protocols with respect to application requirements and available network resources. Performance is increased through application specific tailored protocols and reduced protocol complexity allowing stacks to fit inside the limited memory space of current mobile devices. The framework is written to allow the systems functional requirements to be separated from the non-functional requirements permitting future modules to be incorporated at a later date with little or no disruption to the system. Prioritised multicast media groups provide alternative qualities of service, which allow clients to essentially choose the optimal streaming rate while active proxies perform transcoding and rate-control to allow fine-tuning of service within each group.

1 Streaming Media to Mobile Devices

Mobile communications is a continually growing sector in industry and a wide variety of visual services such as video-on-demand have been created which are limited by low-bandwidth network infrastructures. The distinction between mobile phones and personal device assistants (PDA's) has already become blurred with pervasive computing being the term coined to describe the tendency to integrate computing and communication into everyday life [Dertouzos99, Fong03, Stanford03]. New technologies for connecting devices like wireless communication and high bandwidth networks make the network connections even more heterogeneous. Additionally, the network topology is no longer static, due to the increasing mobility of users. Ubiquitous computing is a term often associated with this type of networking [Tanter02]. The creation of low bit rate standards such as H.263 [H263] allows reasonable quality video through the existing Internet and is an important step in paving the way forward. As these new media services become available the demand for multimedia through mobile devices will invariably increase. Corporations such as Intel do not plan to be left behind. Intel has created a new breed of mobile chip code named Banias. Intel's president and chief operating officer Paul Otellino states that 'eventually every single chip that Intel produces will contain a radio transmitter that handles wireless protocols, which will allow users to move seamlessly among networks. Among our employees this initiative is affectionately referred to as 'radio free Intel'".[1]

Recent research advances in Quality of Service (QoS) have led to numerous solutions to support QoS aware applications, so that the demands for both end system and network resources are met [Hartenstein01, Kassler02]. Two major categories of solution have emerged. First, reservation based systems rely on resource reservation and admission control mechanisms to enforce the delivery of requested QoS to the applications. This approach can often demand major modifications to the design of operation systems and prevalent network protocols. In contrast, adaptation-based systems operate based on best-effort environments, and attempt to adapt themselves or the applications for the purpose of providing the best possible QoS under available resource conditions, and of achieving the most graceful quality degradation in case of scarce resources. It is advantageous to implement such adaptation-based systems in the middleware level, since it does not require tight integration or modifications to the best-effort services in the operating system kernel and network protocol stack, which is the major advantage of adaptation-based systems over reservation-based systems.

---

[1] www.pcplus.co.uk (Article in May 2002 issue)

Products such as Real Audio[1] and IPCast[2] for streaming media are also becoming increasingly common but multimedia, due to its timely nature requires guarantees different in nature with regards to delivery of data from TCP traffic such as HTTP requests. In addition, multimedia applications increase the set of requirements in terms of throughput, end-to-end delay, delay jitter and clock synchronisation. These requirements may not all be directly met by the networks therefore end-system protocols enrich network services to provide the quality of service (QoS) required by applications. It is argued here that traditional monolithic protocols are unable to support the wide range of application requirements on top of current networks (ranging from 9600 baud modems up to gigabit networks) without adding overhead in the form of redundant functionality for numerous combinations of application requirements and network infrastructures.

Figure 0-1: PDAs   Figure 0-2: Mobiles   Figure 0-3: Laptops

In ubiquitous computing, software is used by roaming users interacting with the electronic world through a collection of devices ranging from handhelds such as PDAs (Figure 0-1) and mobile phones (Figure 0-2) to laptops (Figure 0-3). The Java language thanks to its portability and support for code mobility is seen as the best candidate for such settings [Weiser93, Roman02 and Kochnev03]. Flexible and adaptive frameworks are necessary in order to develop distributed multimedia applications in such heterogeneous end-systems and network environments [Schmidt03]. The processing capability differs substantially for many of these devices with PDA's being severely resource constrained in comparison to leading desktop computers [Lin01]. The networks connecting these devices and machines range from GSM, Ethernet LAN, and Ethernet 802.11 to Gigabit Ethernet [Gautney03]. The heterogeneity added by modern smart devices is also characterised by an additional property, which is that many of these devices are typically tailored to distinct purposes [Greenberg99]. Therefore, not only memory and storage capabilities differ widely but

---

[1] www.realaudio.com
[2] www.ipcast.com

local device capabilities, in addition to the availability of resources changing over time (e.g. a Global Positioning Satellite (GPS) system cannot work indoors unless one uses specialised repeaters – see [Jee03]) thus a need exists for middleware to be aware of these pervasive computing properties. With regards to multimedia, applications that use group communication (e.g. video conferencing) mechanisms must be able to scale from small groups with few members, up to groups with thousands of receivers [Tojo03].

The Internet is built on the DARPA protocol suite Transmission Control Protocol/Internet Protocol (TCP/IP) [RFC761], with IPv4 [RFC791] as the enabling infrastructure for higher-level protocols such as TCP and the User Datagram Protocol (UDP) [RFC768]. It can be argued that the protocols underlying the Internet were not designed for the latest cellular type networks with their low bandwidth, high error losses and roaming users, thus many 'fixes' have arisen to solve the problem of efficient data delivery to mobile resource constrained devices [Saber03]. Mobility requires adaptability meaning that systems must be location-aware and situation-aware taking advantage of this information in order to dynamically reconfigure in a distributed fashion [Katz94, Solon03 and Matthur03]. Situations, in which a user moves an end-device and uses information services, can be challenging. In these situations the placement of different co-operating parts is a research challenge. The heterogeneity is not only static but also dynamic as software capabilities, resource availability and resource requirements may change over time. The support system of a nomadic user must distribute, in an appropriate way, the current session among the end-user system, network elements and application servers. In addition, when the execution environment changes in an essential and persistent way, it may be beneficial to reconfigure the co-operating parts. The redistribution or relocation as such is technically quite straightforward but not trivial. On the contrary, the set of rules that the detection of essential and persistent changes is based on is a challenging research issue. This problem is the focus of this book. A middleware framework has been developed that provides uniform access to remote services and device-specific capabilities, the decoupling of the application communications model and the underlying interoperability protocols alongside dynamic extensibility supporting a range of devices from small-embedded systems to full-fledged computers. This opening chapter provides an overview of the subject, outlines the problems in streaming media to mobile devices, depicts some of the traditional solutions and presents proposed solutions for the addressed problem area.

1.1   IP (Internet Protocol)

The Internet Protocol [RFC791] is the basic protocol of the Internet that enables the unreliable delivery of individual packets from one host to another. It makes no guarantees about whether or not the packet will be delivered, how long it will take, or if multiple packets will arrive in the order they were sent. Protocols built on top of this add the notions of connection and reliability. One reason for IP's tremendous success is its simplicity. The fundamental design principle for IP was derived from the "end-to-end argument", which puts "smarts" in the ends of the network - *the source and destination network hosts* - leaving the network "core" dumb. IP routers at intersections throughout the network need do little more than check the destination IP address against a forwarding table to determine the 'next hop' for an IP datagram (where a datagram is the fundamental unit of information passed across the Internet). IP Multicast [Deering99, Almeroth00 and Williamson00] is a bandwidth-conserving technology that reduces traffic by simultaneously delivering a single stream of information to thousands of corporate recipients and homes. Applications that take advantage of multicast include videoconferencing, corporate communications, distance learning, and distribution of software, stock quotes, and news. IP Multicast delivers source traffic to multiple receivers without adding any additional burden on the source or the receivers while using the least network bandwidth of any competing technology. Multicast packets are replicated in the network by routers enabled with Protocol Independent Multicast (PIM) [RFC1112, Fenner00] and other supporting multicast protocols resulting in the most efficient delivery of data to multiple receivers possible. All alternatives require the source to send more than one copy of the data. Some even require the source to send an individual copy to each receiver. If there are thousands of receivers, even low-bandwidth applications benefit from using IP Multicast [Williamson00]. High-bandwidth applications, such as MPEG video, may require a large portion of the available network bandwidth for a single stream. In these applications, the only way to send to more than one receiver simultaneously is by using IP Multicast.

The MBone (Multicast Backbone) is an extension to the Internet to support IP [Deering89, Almeroth00]. The TCP/IP protocol used by the Internet divides messages into packets and sends each packet independently. Packets can travel different routes to their destination, which means that they can arrive in any order and with sizable delays between the first and last packets. In addition, each recipient of the data requires that separate packets be sent from the source to the destination. This works fine for static information, such as text and graphics, but it does not work well for real-time audio and video. Multicasting allows a single packet to have multiple destinations and is not split up until the last possible moment allowing it to pass through several routers before it needs to be divided to reach its final destinations. This leads to much more efficient transmission and also ensures that packets reach multiple destinations at roughly the same time. The MBone consists of known servers (mostly on UNIX workstations) that are equipped to handle the multicast protocol. Tunneling is used to forward multicast packets through routers on the network that don't handle multicasting. An MBone router that is sending a packet to another MBone router through a non-MBone part of the network encapsulates the multicast packet as a unicast packet. The non-MBone routers simply see an ordinary packet. The destination MBone router unencapsulates the unicast packet and forwards it appropriately. The MBone consists of a backbone with a mesh topology which is used by servers that redistribute the multicast in their region in a star topology. The MBone network is intended to be global and includes nodes in Europe. The channel bandwidth for MBone multicasts is 500 kilobits per second and actual traffic is from 100-300 kilobits depending on content. MBone multicasts usually consist of streaming audio and video.

1.2   Wireless Networks

In 1946, the first car-based telephone was set up in St. Louis in the USA. The system used a single radio transmitter on top of a tall building. A single channel was used, and therefore a button was pushed to talk, and released to listen [Tanenbaum03]. This half duplex system is still used by modern day CB-radio systems used by police and taxi operators. In the 60's the system was improved to a two-channel system, called improved mobile telephone system (IMTS). The system could not support many users as frequencies were limited. The problem was solved by the idea of using cells to facilitate the re-use of frequencies. More users can be supported in such a cellular radio system. It was implemented for the first time in the advanced mobile phone system (AMPS). AMPS is an analogue system which is part of first generation cellular radio systems. Second generation systems are digital. In the USA two standards are used for second generation systems - IS-95 (CDMA) and IS-136 (D-AMPS) [Tanenbaum03, Rysavy99]. Europe consolidated on one system called global system for mobile communications (GSM) [Rysavy99] while Japan uses a system called personal digital cellular (PDC).

Type of Network	Bandwidth Latency	Typical Video Performance	Typical Audio Performance
In-Room/Building (Radio Frequency Infrared)	>> 1 Mbps RF: 2-20 Mbps IR: 1-50 Mbps	2-Way, Interactive, Full Frame Rate (Compressed)	High Quality, 16 bit samples, 22 KHz rate

Campus-Area Packet Relay	Approx. 64 kbps	Medium Quality Slow Scan	Medium Quality Reduced Rate
Wide-Area (Cellular, PCS)	19.2 kbps	Video Phone or Freeze Frame	Asynchronous "Voice Mail"
Regional-Area (LEO/VSAT DBS)	Asymmetric Up/Dn 100 bps to 4.8 kbps 12 Mbps	Async Video Playback	Asynchronous "Voice Mail"

Table 0-1: Characteristics of various wireless networks

Wide-area wireless data services have been more of a promise than a reality. It can be argued that success for wireless data depends on the development of a digital communications architecture that integrates and interoperates across regional-area, wide-area, metropolitan-area, campus-area, in-building, and in-room wireless networks. The convergence of two technological developments has made mobile computing a reality. In the last few years, the UK and other developed countries have spent large amounts of money to install and deploy wireless communication facilities. Originally aimed at telephone services (which still account for the majority of usage), the same infrastructure is increasingly used to transfer data. The second development is the continuing reduction in the size of computer hardware, leading to portable computation devices such as laptops, palmtops, or functionally enhanced cell phones. Unlike second-generation cellular networks, future cellular systems will cover an area with a variety of non-homogeneous cells that may overlap. This allows the network operators to tune the system layout to subscriber density and subscribed services. Cells of different sizes will offer widely varying bandwidths: very high bandwidths with low error rates in pico-cells, very low bandwidths with higher error rates in macro-cells as illustrated in Table 0-1. Again, depending on the current location, the sets of available services might also differ.

1.3 Distributed Multimedia

The term media refers to the storage, transmission, interchange, presentation, representation and perception of different information types (data types) such as text, graphics, voice, audio and video while the term multimedia can denote the property of handling a variety of representation media in an integrated manner. It is necessary for a multimedia system to support a variety of representation media types. The range of media types could be as modest as text and graphics or as rich as animation, audio and video. This alone is not sufficient for a multimedia environment. It is also important that the various sources of

6

media types are integrated into a system framework [Wolcott01]. Note that it is important to draw a distinction between multimedia systems, which operate on a single computer workstation, and those that can span a networked environment. The term distributed multimedia is introduced to describe the general case of a number of multimedia workstations interconnected by one or more multi-service networks. In addition, a distributed multimedia is defined as an application, which runs over a distributed multimedia system. The problems of managing multimedia in a distributed system are great and introduce a number of unresolved research problems. In contrast, the technological problems of standalone multimedia workstations are much better understood. Much functionality used for some time in general-purpose computing can be adapted for specific use in distributed multimedia systems. Two domains in which this trend is most visible are communication protocols and operating systems. In order to benefit from the services, multimedia application developers must rely on simple application programming interfaces (APIs). Unlike traditional computer systems characterised by short-lived connections that are bursty in nature, Streaming Audio/Video sessions are typically long lived (the length of a presentation) and require continuous transfer of data. Streaming services will require, by today's standards, the delivery of enormous volumes of data to customer homes. For examples, entertainment NTSC video compressed using the MPEG standards requires bandwidths between 1.5 and 6 Mbits/s. Many signalling schemes have been developed that can deliver data at this rate to homes over existing communications links [Starr99]. Some signalling schemes suitable for high-speed video delivery are:

ADSL: The Asymmetrical Digital Subscriber Loop (ADSL) [Sari99, Bingham00] takes advantage of the advances in coding to provide a customer with a downstream 1.536 Mbits/s wideband signal, an upstream 16 kb/s control channel and a basic-rate ISDN channel, on existing twisted copper-pair. The cost to the end-user is quite low in this scheme, as it requires little change to the existing equipment.

HDSL: The High-Speed Digital Subscriber Line (HDSL) [Goralski98, Ives01] allows transmission of up to 800 kb/s up to 5.5 km on existing copper lines. With two such circuits in parallel, the technology can support up to 1.544 Mbits/s full duplex communication.

CATV: Cable TV (CATV) [Yates90, Forouzan02] uses a broadband coaxial cable system and can support multiple MPEG compressed video streams. CATV has enormous bandwidth capability and can support hundreds of simultaneous connections. Furthermore, as cable is quite widely deployed, the cost of supporting Video-on-demand and other services is significantly lower. It also requires adaptation to allow bi-directional signalling in the support of interactive services.

ADSL is currently the prime candidate for supporting interactive multimedia in UK homes due to its predicted wide deployment [Minoli02]. ISDN [Stallings98] is a candidate; but at the basic rate (128 kb/s) it has insufficient bandwidth to support entertainment-quality video delivery but enough for Internet standard video conferencing applications.

1.3.1   Multimedia Sychronisation

A major requirement of distributed multimedia applications is the need to provide comprehensive support for real-time synchronization. Many existing distributed systems platforms provide extensive synchronization primitives in terms of, for example, semaphores, mutexes, condition variables or message passing services. It is less common to support real-time requirements directly. Two styles of real-time synchronization can be identified in distributed multimedia applications. These are:

1. Intra-media synchronisation is required to ensure that the real-time integrity of a particular continuous media type is preserved in an interaction. For example, this style of synchronisation would be required to ensure that a video stream is presented with the required throughput, jitter and latency characteristics. This corresponds directly to quality of service requirements for continuous media as discussed above. Indeed, quality of service management can be viewed as a mechanism to achieve intra-media synchronisation

2. Inter-media synchronisation is more complex and is concerned with arbitrary real-time relationships between different interactions (discrete or continuous). Such relationships can either be expressed in absolute terms with respect to a global real-time clock (for example, two stream interactions should start at a given time) or in relative terms with respect to each other (e.g. a particular video clip should start 100 milliseconds after the end of an animation).

Examples of inter-media synchronization include lip synchronization between audio and video channels, synchronization of stereo audio channels and synchronization of text subtitles and video sequences. The first two inter-media examples refer to constraints between continuous media types; the third example illustrates inter-media synchronization between a continuous media type and a discrete type.

1.3.2   Operating Systems support for Multimedia

Another problem well recognised by the multimedia community is the limited value of popular operating systems, again due to the real-time nature of continuous media. Decent multimedia presentations depend on the ubiquitous availability of the necessary resources. This can be easily verified by observing the jitter and slowdown of a presentation, caused when other jobs are initiated. These problems are not due to a real shortage of resources, but to inappropriate scheduling. Continuous media processing is known to be of 'soft real-time' nature, where occasional computing errors or failures to meet deadlines (such as missing or delivering too early or delaying a video frame) do not result in failure. Further differences include the fact that time-critical processing of continuous media is mostly periodic (e.g. reception of video frames), and that applications may adapt to variations of resource availability via QoS negotiation (e.g. by decreasing the quality of video playback). The principle requirements placed on multimedia-capable operating systems are *QoS*-based resource management; *Real-time CPU scheduling*. Rate-monotonic and earliest deadline-first are the most frequently used policies; *Memory, buffer and file management* policies to support real-time scheduling; *Support for real-time synchronisation*. This implies efficient inter-process communication, even notification and mutual exclusion mechanisms and *Low overhead task management*; this is needed due to frequent task switching. Lightweight threads and thread switching in user space are employed in some systems.

1.3.3   Multimedia Compression

Multimedia compression is concerned with reducing the amount of data required to reproduce an output stream such as digital video or audio. It is a key component in facilitating the widespread use of streaming media, which is currently prevented by the mismatch between the huge storage and transmission bandwidth requirements of video and the limited capacity of existing computer systems and communications networks. Although the range of video compression standards now available, such as H.261 or MPEG-2, exhibit acceptable performance when operating at high bit rates (eg >64 Kbps), they perform badly at very low bit rates, prompting the need for the development of the next generation of video compression standards. MPEG/audio is a popular standard for audio-only applications to compress high fidelity audio data at much lower bit rates. While the MPEG/audio compression algorithm is lossy, often it can provide "transparent", perceptually lossless, compression even with compression factors of 6-to-1 or more. The algorithm works by exploiting the perceptual properties of the human auditory system

[Pan95, Feamster99].

	Sample Rates	Bandwidth Requirements
Telephone Speech	8 kHz, 16 bit, 1 channel	128 kbps
CD Quality Sound	44.1 kHz, 16 bit, 2 channels	1.41 Mbps
Standard TV Video	640x480 pixels x 16 bit x25 fps	122 Mbps

Table 0-2: Uncompressed multimedia bandwidth requirements

Uncompressed multimedia can require large amounts of storage capacity and high bandwidth. For instance, the following picture with the dimensions 140 x 90 mm and a resolution of 150 dpi (dots per inch) would require approximately 3 MB of memory. Computational requirements for various compression methods used to reduce the volume of data are shown in Table 0-2.

Audio Compression	Sampling Rate	Bits per sample	Bits	Computational
G.721	8 kHz	8 bits	32 bits	Low
G.723	8 kHz	8 bits	24 & 40 bits	Low
DVI	8 kHz	8 bits	32 bits	Low
GSM	8 kHz	4 bits	13.2 bits	Low
MPEG-1 Layer 1	32, 44 & 48 kHz	16 bits	32-384 bits	High
MPEG-1 Layer 2	32, 44 & 48 kHz	16 bits	32-384 bits	High
MPEG-1 Layer 3	32, 44 & 48 kHz	16 bits	32-320 bits	High

Table 0-3: Audio compression algorithms [Pranata02]

Video Compression	Resolution Format	Bit Rate	Computational
H.261	CIF, QCIF	64 kbps	Low
H.263	CIF, QCIF, SQCIF, 16CIF	28.8-768 kbps	Low
MPEG-1	352x240 pixels, 30fps	1.5 Mbps	High
MPEG-1	352x288 pixels, 25 fps	1.5 Mbps	High
MPEG-2	720x480 pixels, 30fps	1.5 Mbps	Very High
MPEG-2	1920x1080 pixels, 30fps	1.5 Mbps s	Very High
MPEG-4	1920x1080 pixels, 30fps	28.8-500 kbps	High

Table 0-4: Video compression algorithms [Pranata02]

Compression of media is essential when transmitting media over a wide area network. This involves the source using a media encoder to compress (encode) the media and the receiver using a corresponding decoder to uncompress (decode) the media but this places an extra burden on both communicating devices CPU/processing resources. The computation load ascribed to various compression algorithms can be seen in Table 0-3 and Table 0-4.

1.4   Multi-Layered Multimedia Transmission

Scaling means to sub sample a data stream and only present some fraction of the original content. Scaling can be done at either the source or the sink of a stream. Frame rate reduction, for example, is usually performed at the source whereas hierarchical decoding is a typical scaling method applied by the sink. It is useful to scale a data stream before it enters a system bottleneck otherwise it is likely to contribute to the overload of the bottleneck resource. Scaling at the source is usually the best solution as there is no need for transmitting data in the first place if it will be discarded within the system. Media scaling is based on the assumptions that it is possible to sub sample a data stream and only present some fraction of its original contents and that mechanisms are available that are able to detect the data traffic variations over the network (e.g. network congestion). When network congestion is detected by monitoring functions, the stream is scaled down so that it requires less bandwidth. Clients therefore receive a lower quality sub sample of the stream. This is preferable to attempting to deliver the whole stream with the likely result of receiving no data at all and further contributing to the congestion. The stream can be scaled up to fuller quality later on once congestion has ceased. Applications for scalable flows can be classified into the following areas:

1.4.1   Congestion Control

Applications may prioritise sub-flows, and then discard according to priority when congestion is encountered at a node so as to reduce packet loss achieving a level of reduction in quality rather than larger packet losses. Nodes decide when a packet is to be discarded and is generally a function of the local scheduling mechanisms [Campbell97b, Bajaj98]. Generally, congestion control applications require use of hierarchical encoding methods.

1.4.2   Bandwidth And Admission Control

Bandwidth and admission control attempts to trim the flow to fit the limited resources on the network. The Internet contains no explicit mechanisms for admission control and resource reservation [Cetin01]. Scalable flows can prove to be a useful tool for bandwidth control in these environments. Network managers often know a priori that resource choke-points exist in the network therefore filters can be instantiated manually or via some sort of out-of-channel mechanisms local to a node. Generally, either simulcast or hierarchical encoding can be used for this class of application, as filter criteria are stable over relatively long time-scales.

1.4.3   Traffic Selection

Scalable flows can be used to provide mechanisms for the receiver to explicitly control the type of traffic directed to it such as electing to receive only the low-resolution portion of a flow while a video window is thumbnail size, and all the portions of a flow when it is full size. Filter propagation mechanisms can be used to push this selection further on to the network, reducing the consumed bandwidth. Simulcast or hierarchical encoding can be used for this class of application. Multicasting the same media stream to receivers distributed in a wide range may congest certain low-capacity or heavily loaded areas of the network, while some other high capacity or lightly loaded areas remain under-utilised. A fair mechanism [Cheung96, Bradshaw01] should ensure that each client receives a stream with a quality commensurate with its processing power and bandwidth capacity on its connected link. One solution is to multicast different layers of video containing progressive enhancements to different multicast addresses where clients decide which multicast group(s) to subscribe to [McCanne96, Gopalakrishnan00].

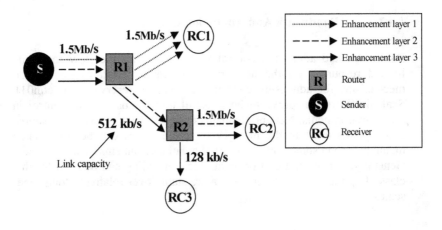

Figure 0-4 : Example of multi-layer transmission

The minimal amount of data needed for an acceptable representation of the original data stream is contained in a base layer with each additional layer providing a Quality improvement. Figure 0-4 depicts an example of a multi-layered transmission approach where the sender S transmits its data split into three streams over three layers. There is sufficient bandwidth available on the link between the router R1 and receiver RC1; therefore RC1 subscribes to all three layers. The link between the router connected to RC2 and RC3 has a lower capacity so only the two lower layers of the three can be forwarded. The available bandwidth on the link to receiver RC3 is even more restricted; therefore RC3 decides only to receive the base layer. The link to RC2 has a capacity of 1.5 MBps, but the capacity between the two routers is restricted to 512kbit/s, therefore RC2 can only subscribe to two layers, i.e. the base layer and enhancement layer 1. The different approaches introduced in the literature for utilising layered data transmissions can roughly be divided into two categories: cumulative and independent layering.

1.4.4 Cumulative Layering of Media

In the Cumulative Layered case [Legout00], each layer provides refinement information to the previous layers. MPEG-2 standard offers four different 'scalability modes' [Campbell95, Gopalakrishnan00]: spatial, SNR, and temporal scalability as well as data partitioning. Up to two different scalability modes (except from data partitioning) in any combination can be used simultaneously, resulting in a 3-level representation of the signal. In spatial scalability, the base and enhancement layers operate at different spatial resolutions. In SNR scalability, both adjacent layers have the same resolution and the enhancement refines the quantization process performed in the base layer. Data partitioning splits encoded data into two bit streams, one containing more and the other containing less important data. Data partitioning and SNR scalability lead to a drift problem caused by motion compensated coding if only the lower bit rate can be received. Another problem with cumulative layering is that of resynchronisation as the different layers may have been delayed en-route due to different routes. In [Hoffman96], the authors describe a scheme that combines hierarchical data transmission and resource reservation where the different data layers are transmitted over different multicast sessions. End systems wishing to receive a certain data layer, join the multicast group on which the data layer is being sent and issue a RSVP reservation request based on the characteristic of that data layer [Zhang93, Raspall01]. If the reservation request cannot be granted due to lack of resources, the receiver must leave the session. It is usually difficult to know the exact characteristics of a certain stream in advance and there would be a tendency to over reserve resources in order to guarantee the requested QoS level. Reservation can only be used if the network nodes involved support it, which currently is not the case.

Receiver-Driven Layered Multicast (RLM) [McCanne96] is a receiver-based rate adaptation technique, which combines a layered source-coding algorithm with layered multicast transmission. Data sources adopt passive roles by simply transmitting each layer of data on a separate multicast group and leave the adaptation to receivers, which join and leave multicast groups. Receivers on detecting congestion drop layers and add layers on detecting spare capacity [Raspall01]. To find out if the current level of subscription is too low so-called 'join experiments' are performed where layers are added. If this results in congestion, the new layer is dropped quickly and a timer is set for a new experiment otherwise the layer is kept. Adaptive Layered Transmission (ALT) [Sisalem98, Kassler01a] seeks to improve upon RLM, by having the sender dynamically redistribute a data stream on the different layers based on the feedback of the receivers that allows for the dynamic discovery of the appropriate transmission rate for each layer.

1.4.5   Independent Layering of Media

The Independent Layered Data Transmission approach multicasts replicated streams [Cheung94/95/96, Li96, Gopalakrishnan99, Anastasiadis02] for improving fairness of video distribution in a heterogeneous network with a small overhead in terms of bandwidth cost. This scheme is often called simulcast because the source transmits multiple copies of the same signal simultaneously at different rates resulting in different qualities. As the streams contain all the necessary information for decompression, the receivers need only join one multicast session. This method avoids the resynchronisation and drift problems seen with cumulative approaches and is the approach adopted by this research. Independent layered data techniques require each subflow to be independent therefore a packet loss in one subflow should have no impact on the other subflows. In contrast, because of the inter-dependence of data streams within hierarchical coding, the loss of a packet in one subflow can have a greater impact on the other subflows [Anastasiadis02]. If a packet loss were to occur in a high quality data stream, there would be a quality loss for that particular data unit. In the case where the packet is aligned to a slice boundary, the slice corresponding to a row of macro blocks, only a small strip of the image would experience a quality loss. If the high priority subflow were to experience packet loss, no picture reconstruction at all would be possible.

1.5   Application Controlled Quality of Service

A need exists for sharing of the network between data streams carrying differing payloads with dissimilar QoS requirements. Networking transport technologies such as Asynchronous Transfer Mode (ATM) [Cuthbert93] have various categories of guarantees ranging from best effort (attempts to reach destination are not guaranteed) to guaranteed (system has failed if it fails to deliver all data). The Internet is composed of heterogeneous network technologies offering varying services therefore, the guarantee required for multimedia applications such as video conferencing is impractical on a network such as the Internet at present [Cobley98, Kassler02].

A proposal for applying QoS control on the Internet is the Resource reSerVation Protocol (RSVP) [Zhang94, Kassler02, Liebl01] defined by the Internet Engineering Task Force (IETF). RSVP is part of an effort for enabling integrated service on the Internet: the ISA (Integrated Service Architecture) [Wang00]. RSVP is based on flow control and receiver initiated reservation by allowing priorities to be set in modified routers along the data path over the Internet, which basically establish, maintain and release reservations. These mechanisms have focussed on regulating competition for network resources at the network level involving various aspects in resource reservation and allocation, as well as flow control, admission control and negotiation protocols. One promising mechanism is for the customer to set the type-of-service (TOS) bits in the IP header to indicate the level of QoS the particular packet requires. The TOS bits can be set to indicate that the information is real time (e.g. VoIP) or requires a set QoS (e.g. bronze, silver or gold). The TOS information is used by the access router to ensure that the customers date is within the agreed contract and then to place information with similar QoS requirements in similar output queues. The basic idea is that high priority/low latency information is given the highest precedence and hence gets priority for access to the transmission line. One technique that is multi-protocol label switching (MPLS) [MPLS00, MPLSforum03] or Cisco's proprietary tag implementation [Cisco00, Ubik03].

Systems introducing adaptation behaviour into the application at the end-system (and throughout the middleware) are generally referred to as Application Controlled QoS systems, beyond the control mechanisms of the transport layer. This perspective is different from the resource reservation approach in that it mainly deals with protocols and techniques located and operated in the middleware architecture and end-systems, rather than intermediate network switches. The advantage of this approach over the previous one is that it does not need to drastically change existing protocols running in current networks, the implication being that the QoS can be implemented with minimal modifications. The traditional relation between service user and service provider is a simple contract with applications specifying their requirements by a target value [Cingser93, Cheriton95, Corman01] or target range [Clark97] for QoS parameters. Service providers offer QoS in a best-effort manner such as OSI TP4 [Huitema90] or in a guaranteed manner such as ST-II [Mitzel94]. Generally, applications have only a very limited knowledge on available resources, network services, and end system load, and they may change drastically [Ferrari90, Danthine93]. Furthermore, applications want to attain more than one (possibly contradicting) objective such as high performance and low costs [Campbell98]. The adaptation behaviour that is currently designed for the applications is usually built into the application such as the video player and the multimedia play-out control algorithm is implemented inside each application, thus the application can maintain a certain expected play-out level for the client when adapting to fluctuations in the delivery of data.

The goal of multicast control is for the sender to ideally choose its rate to match the maximum capability of each receiver, while at the same time minimising the amount of bandwidth used. The solution is non trivial as there must be a balance between the sender transmitting too high a rate for hosts with fewer resources and the sender transmitting at too low at rate which leads to higher capability hosts performing at sub-optimal levels. Approaches proposed to handle the real-time aspect of video distribution over networks are mainly divided into the use of a network capable of resource reservation to provide performance guarantees [Ferrari94, , Shacam94, Bettati95] and the use of adaptive control to adjust multimedia traffic characteristics to meet the current network's capacities [Bolot94, Cheung96, Li97]. This latter approach is more compatible with the current architecture and capabilities of the Internet as technologies such as RSVP and ATM, which offer network-level reservations, are far from being used ubiquitously for real-time video distribution [Zhang93]. Even when reservations are available it can be difficult to make accurate reservations so some adaptation is required to allow tolerance in reservation accuracy.

1.6   Layered Protocol Stacks

In the vocabulary of design patterns [Gamma95, Coad95, Buschmann96] a layered protocol stack is familiar as a chain of responsibility. Layered protocol stacks have appeared in the X-Kernel project [Peterson96a, Peterson96b] and Horus [Renesse96] and commercial systems such as Orbix [1]. The pioneering work was introduced in Ritchie's streams for Unix [Ritchie90]. Modern networks offer end-to-end connectivity but the offered services may not fulfil the requirements of distributed applications and must therefore be enriched by an end-to-end communication support. This leads to a communication system consisting of 3 layers, as illustrated in Figure 0-5. End systems inter-communicate through layer T, the transport infrastructure. The service of layer T is a generic service corresponding to layer 2, 3 or 4 services in the OSI Reference Model. In layer C the end-to-end communication support adds functionality to the services in layer T. This allows the provision of services at the layer A for distributed applications (A-C Interface). Layer C is decomposed into protocol functions, which encapsulate typical protocol tasks such as error and flow control, encryption and decryption, presentation coding and decoding among others. A protocol graph is an abstract protocol specification, where independence between protocol functions is expressed in the protocol graph. If multiple T services can be used, there is one protocol graph for each T service to realise a layer C service. Protocol functions (modules) can be accomplished in multiple ways, by different protocol mechanisms, as software or hardware solutions [Plagemann93] with each protocol configuration in a protocol graph being instantiated by one of its modules [Plagemann95].

---

[1] http://www.iona.com

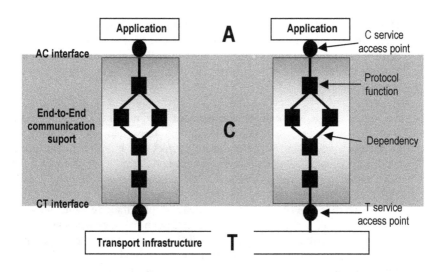

Figure 0-5 : 3 layer model [Plagemann95]

Layering is a form of information hiding where a lower layer presents only a service interface to an upper layer, hiding the details of how it provides the service. Information hiding has numerous benefits but it can sometimes lead to poor performance especially in cases where optimisations are possible given knowledge of what the lower layers are doing [Clark90]. A flow control protocol that is responsible for throttling a source when it thinks that the network is overloaded may assume that overloads are correlated with packet loss thus that protocol could throttle the source-sending rate when it detects a packet loss (TCP). A flow control protocol might not know how packets are transferred across the network if the flow control protocol were layered above a protocol that is actually responsible for data transfer. The end system may be connected to a network over a lossy wireless link so that packet losses are mostly due to link errors rather than network overload. Unfortunately, even in this situation, the flow control protocol thinks that the network is congested and throttles a source even when there is no need to do so. The information that a packet is lost on the link, which is available to the lower layer, is hidden from the higher (flow control) layer, which results in inefficient flow control. A solution would be for the lower layers to inform the upper layer concerning packet loss so that the flow control layer could distinguish between the link and congestive losses and perform a better job. This violates layering, as the flow control layer now knows about the details of data transfer over its local link so if the data transfer layer changes because the end system is using a different link technology (e.g. Wireless link versus Wired link), the flow control layer, which ought to be independent of the link technology, must also change.

18

This illustrates the tension between information hiding on one hand and performance on the other hand. Intelligent protocol stack designs should 'leak' enough information between layers to allow efficient performance, but not so much that it is difficult to change the implementation of a layer.

As layers are self-contained small pieces of functionality, independent from other layers and relatively small; they can be verified for correctness more easily than large chunks of interdependent code. Layering forces better structuring; as code for a certain type of functionality is localised in one place and layers can be easily replaced/upgraded with new versions. A protocol layer typically either modifies a message (e.g. by adding a header), or it may delay its delivery, (e.g. to preserve ordering in a FIFO layer). Each message has a layer type tag, describing the layer from which it originated. Layers also generate messages stamped with its unique type. Each layer checks whether a message's layer type matches its own and if so, the layer processes the message otherwise it will be passed on to the next layer. The protocol layer interface has methods to process messages from layers above or below (Down or Up) which are called when a message passes through the layer. Most protocols allow for some variation in packet sizes, timeouts, among others; but the defaults provided may not be optimal for all situations. TCP uses window sizes from 1 to 7, and packet sizes in powers of 2 ranging from 64 through 4096. If over a noisy link it drops more than 10 percent of all packets, then losses may be reduced by lowering the packet size and shrinking the window. On non-lossy links, the overhead of sending ACKs for every 128 bytes may prove wasteful, therefore the packet size could be increased to 512 or whatever is closer to the maximum packet size. Interchanging ACK and NAK schemes may also lead to improved performance as in large scale multicast reliable communications [Yamamoto97]. When the ACK-based scheme is used, an ACK-implosion problem can occur, as the number of participating receivers is large. In the NAK-based approach, a NAK-suppression mechanism [Floyd97] can be applied to improve its scalability. The NAK-based approach with the NAK-suppression mechanism is suitable for scalable multicast communications [Yamamoto01].

A flexible protocol system allows the dynamic selection, configuration and reconfiguration of protocol modules to dynamically shape the functionality of a protocol in order to satisfy application requirements or adapt to changing service properties of the underlying network. Flexible end-to-end protocols are configured to include only the necessary functionality required to satisfy an application QoS requirements for the particular connection. Some uses that dynamic stacks may be used for include increasing throughput where environmental conditions are analysed and heuristics are applied to decide if change would bring about optimal performance; interoperability in that dynamic stacks can simplify the interoperability process, by allowing code for protocol stacks to be written once and placed on repositories where they can be downloaded onto end systems so they can adopt the same stack; Security can be increased at run-time, for example, when an intrusion detection system dynamically responds to unusual behaviour and robustness, where faulty components can be detected and replaced to improve robustness (e.g. a mobile computer connected to an Ethernet LAN may automatically detect that its wired connection is broken forcing a switch to a Wireless LAN or GSM connection. Here it is profitable to dynamically load a new stack module optimised for the different characteristics of wireless connections).

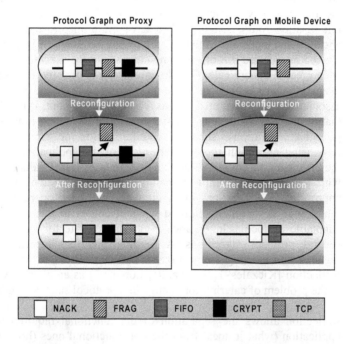

Figure 0-6 : Streaming Video Application Object Graph

Figure 0-6 depicts a typical layered protocol graph. When a network device is switched from e.g., a low speed cellular network to a high speed wireless LAN, a session manager will detect that the underlying network has changed. The protocol stack control mechanism will install a new stack module, (e.g. which might contain an algorithm for conversion of a colour video stream to a monochrome video stream). As the framework

20

separates the configuration of objects from the implementation of the objects, it makes it easier to build adaptive applications.

1.7 Dynamic Stack Configuration

Programs are frequently developed in one environment, and later installed and configured in a different target environment where much of the configuration activity relates to setting program variables (e.g. IP addresses, server names, etc) to appropriate values. Environment variables or text files offer a primitive solution; but as programs increase in complexity, configuration possibilities also increase. Therefore to support program flexibility, programming languages must frequently control configuration. One solution is to incorporate an interpreter into the run-time environment so program configuration files can contain initialisation routines, which then have the ability to use conditionals, abstractions, loops etc. of that programming language. A language can be classed interpreted if it offers some mechanism for execution of blocks of code created dynamically (i.e. the interpreter is directly accessible to the language). Java is an example of an interpreted language. One weakness associated with interpreted languages is efficiency but, it can be argued that in many systems, a substantial amount of execution time is spent on primitive operations, such as communication or input/output calls, possessing execution times independent of the application coding thus the efficiency of compiled languages here is of lesser value. Inheritance with regards to the implementation of distributed applications is an appropriate tool (e.g. passing arguments, call-back operations etc) but inheritance is inevitably limited when it comes to the composition of adaptable protocols. Inheritance is not appropriate for choosing among several protocol algorithms at runtime therefore incorporating new features into existing applications is difficult. These limitations lead to Reflection [Kiczales91, Sullivan01, Lorenz03] as an alternative solution to the problem of catering for change in a protocol stack at runtime.

Reflection allows the separation of the functional requirements of an application (what it does) from the non-functional ones (how it does it). It is based on the Meta-object Protocol (MOP) defined by Maes [Maes87, Arcangeli00]. The Java Reflection API puts particular emphasis on the use of Reflection in distributed systems by taking advantage of the network-centric capabilities of Java. The aim is to overcome the limitations of the black box approach to software development and to open up key aspects of a framework to the application. There should be a principled division between the functionality provided (base interface) and the underlying implementation (Meta Interface) [Rao91, Bruneton00]. Reflective systems by nature support inspection, and adaptation. Reflection provides disciplined, object-oriented access to the implementation of the functional abstraction, which allows the application developer to alter the behaviour of the application by choosing the implementation of a component best suited to the application. Metadata that describe an implementation are specified separately from the functional aspects of the implementation, allowing them to be changed in a simpler manner thus allowing a system to reason about itself and change its behaviour based on both self-knowledge and current properties of the system. An object developed using reflective techniques is more generic because it can handle a larger set of system properties through support for multiple implementations.

In a reflective system, the meta-level interface provides operations to manipulate a self-representation of the underlying implementation. A self representation is where internal structures and their represented domain are linked in such a way that changes in one domain cause a corresponding effect on the other. The beauty of this paradigm is that the self-representation always provides an accurate representation of the system, and ultimately can bring modifications or extensions to itself by virtue of its own computation. Reflection allows applications to adapt the internal system behaviour by modifying the behaviour of an existing service (e.g. tuning the implementation of a protocol to operate more efficiently over a wireless link), or dynamically reconfiguring the system (e.g. inserting a filter object to perform transformation of a video stream) [Blair97, Marangozova00]. One may feel that inheritance could be used to implement a new application class, based on an original application class, which implements some extra non-functional code required by the application. This is true in some cases; but what follows are some of the reasons reflection can be favoured over inheritance [Tassel97].

- Inheritance is static whereas meta-objects can be changed at run time thus changing the extra functionality that the application class receives, but inheritance is static and therefore code re-writing and re-compiling would be required to obtain equivalent changes.

- There are no selector facilities in inheritance but a scheme could be designed based on using one of the calling method parameters as a selector.

- When designing or implementing the meta-object, no access to the application code is required whereas with inheritance the application class source code is required.

Java provides natural support for run-time interface discovery as a Java class file contains explicit data about the names, visibility and signatures of the class and its fields and methods. This was originally provided to enable late loading but the runtime environment also offers this information for other purposes [Redmond00]. Java 1.1 provides a reflection API as a standard facility, allowing any authorized class to dynamically find out and access the class information. The java.lang.Class class has been greatly enhanced in Java 1.1, which now includes methods that return the fields, methods, and constructors defined by a class. These items are returned as objects of type Field, Method, and Constructor, respectively. These new classes are part of the new java.lang.reflect package, and they each provide methods to obtain complete information about the field, method, or constructor they represent, but it still falls short of providing the ability to incorporate run-time behavioural reflection necessary for inserting modules at run-time [Vayssière00, Watanabe00].

1.8 Distributed Messaging Middleware

Back in the 1960s the introduction of networked systems consisted of monolithic systems and mainframes. This architecture consisted of dumb terminals all connected to the mainframe or central control. The advent of the PC made possible a dramatic paradigm shift from the monolithic architecture of mainframe-based applications, which revolutionised the computing networking industry with the introduction of the two-tier client/server. This involved the offloading of some processing from the mainframe to the client. The two-tier architecture evolved into a multi-tier Client/Server architecture with the most popular of the multi-tier architecture being the three-tier model. This was typically partitioned logically into the user interface layer, the business rules layer, and the database access layer [Bolton01]. The next generation introduced a more versatile Distributed Systems model. This architecture takes the concept of the multi-tier model one step further by simply encapsulating the data and business logic within an object and allowing it to be located anywhere within a network. This distributed system architecture offers a much more efficient, flexible and shared approach consisting of individual components. A Distributed object-oriented system is one whose objects are all logically part of a single domain model, yet physically may be spread across a network of machines. Distributed systems have always been important, but they have become even more pervasive over time, given the economics of hardware and networks, and the increasing globalisation of many businesses [Sybex02]. The Internet and cheaper availability of high-performance PCs and networks have created the opportunity for the development of proficient Distributed Systems in recent years. Such systems may comprise of the multi-tiered approach that consists of a Client side, Middle-tier, and a Server side.

Messaging middleware is an inter-connection middle-tier solution that simplifies the technical challenges and reduces the time and effort required in sharing of information and processes. Messaging middleware architectures at their simplest provide interfaces between applications, allowing them to send data back and forth to each other synchronously and/or asynchronously. Messaging middleware and e-mail messaging systems provide similar functionality. The primary difference is that messaging middleware deals with transactions between programs, whereas e-mail messaging deals with messages between people. This structure of middleware can be classified as a Messaging Oriented Middleware (MOM) [Loyall01].

Message-oriented middleware offers many advantages over other such middleware technologies such as Remote Procedural Call (RPC) [RFC 1831] including time independence of components, where the message sender and recipient do not have to be online at the same time, since MOM queues messages when their recipients are not available and location independence of components, which permits the ability to transfer either sender or receiver from one computer to another without bringing the system down due to its communication via topics (or channels). Message-oriented middleware also provides better scalability with the need to only send one single message to a large number of receivers. RPC enforces the sending of individual message to each of the requesting receivers [Bershad90]. CORBA [OMG02a], RMI [Javasoft02] and DCOM [Eddon98] are all based on the RPC model.

1.8.1   CORBA and DCOM

Common Object Request Broker Architecture (CORBA) is an object-oriented architecture, and a language-independent as well as vendor independent architecture. The Object Management Group (OMG) is a consortium of hardware, software and end-user companies formed to create the CORBA architecture [OMG02b]. CORBA specifies the architecture of an Object Request Broker (ORB) regulating interoperability between objects and applications. CORBA is used in the development of distributed applications that provides a standard mechanism for defining the interfaces between components [Rosenberger98]. The Object Request Broker (ORB) is the software that manages communication between objects and is fundamental to the CORBA architecture (See Figure 0-7). An ORB is a software component whose purpose is to facilitate communication between objects. It does so by providing a number of capabilities, one of which is to locate a remote object, given an object reference. Another service provided by the ORB is the marshalling of parameters and return values to and from remote method invocations. As part of the CORBA architecture the Interface Definition Language (IDL) specifies interfaces between CORBA objects, which is the key to ensuring CORBA's language independence. A client application can be written in C++ while communicating with a server written in Java via the CORBA architecture [Rosenberger98].

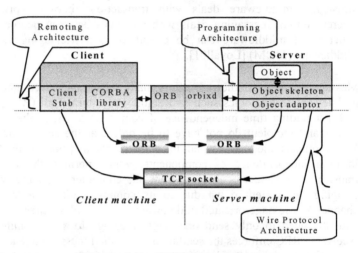

Figure 0-7 : CORBA Architecture

The CORBA specification does not address implementation details and leaves many areas undefined [Gokhale02a]. This omission unfortunately has resulted in proprietary technologies used by the various CORBA vendors [Gokhale02b]. The CORBA model is lacking for the class of systems requiring timely guarantees [DiPippo97]. It is fundamentally based on a blocked synchronous RPC model [Resnick96, Wrox03], rather than an asynchronous Message-Oriented Middleware (MOM) model, which hinders the creation of real-time systems. One may build real-time requests using one-way CORBA operations but some CORBA implementations execute subsequent one-way operations in a LIFO fashion [Resnick96]. These implementations required yet another sequencing protocol and introduced considerable jitter into the video stream, confirming the notion that applying the CORBA request/reply model is not appropriate for streaming time-dependent data transfer. CORBA calls are not asynchronous in reality. The CORBA specification even permits the ORB to block while sending a one-way call and many of the leading ORB vendors have implemented their one-way calls in a blocked manner [Resnick99]. Such behaviour is highly undesirable when high-speed video is to be transmitted and displayed. The OMG special interest group on Real-Time CORBA is studying the issues involved with real-time processing in CORBA but there is no concrete specification from OMG yet. Schmidt, et. Al. [Schmidt99], have developed an architecture for real-time additions to CORBA. They discuss optimisations for high performance, the development of a real-time inter-orb protocol and real-time scheduling.

DCOM

The Distributed Component Object Model (DCOM) [Thai99] was Microsoft's entry into the distributed computing foray, offers capabilities similar to CORBA. DCOM is a relatively robust object model that enjoys particularly good support on Microsoft operating systems because it is integrated with versions of Windows since Windows 95. The availability of DCOM is sparse outside the realm of Windows operating systems. Microsoft is working to correct this disparity by partnering with third party companies to provide DCOM on platforms other than Windows. Yet the fact remains that most organizations have a multi-operating system environment, and although Windows XP seems likely to become dominant over the next few years, diversity will never completely go away. DCOM is an enhancement of COM [Gordon00, Ewald02, Templeman03], which allows the communication of COM processes over a network supporting the creation of a distributed system (See Figure 0-8).

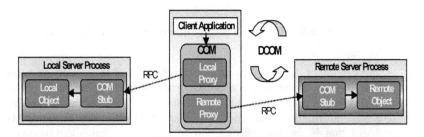

Figure 0-8 : DCOM interconnection of the Client & Server

COM and DCOM are best considered as a single technology providing a range of services for component interaction, from services promoting component integration on a single platform, to component interaction across heterogeneous networks. COM and the DCOM extensions are merged into a single run time providing both local and remote access.

26

ActiveX components are reusable and can be seen as the building blocks for providing the 'glue' that connects components together based on COM and DCOM model [Gordon00]. ActiveX is the standard component model developed by Microsoft and based on a subset of Microsoft's older OLE specification [Nallet00].

1.8.2   Java RMI

Java RMI is part of the Java Development Kit (JDK) [Javasoft02], which provides the core communication layer for any Java object wishing to communicate over an easy-to-use distributed network. RMI applications are often comprised of two separate programs: a server and a client. A typical server application creates some remote objects, makes references to make them accessible, and waits for clients to invoke methods on these remote objects (See Figure 0-9).

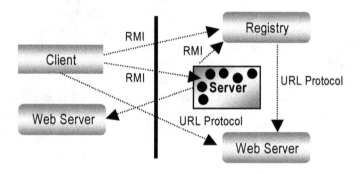

Figure 0-9 : RMI interconnection of the Client & Server

A typical client application gets a remote reference to one or more remote objects in the server and then invokes methods on them. RMI provides the mechanism by which the server and the client communicate and pass information back and forth. Java RMI runs atop of the JRMP protocol that is for the construction of pure Java networked systems. Currently, Java RMI uses a combination of Java serialisation and the Java Remote Method Protocol (JRMP) to turn standard method invocations into remote method invocations.

RMI it can be argued is significantly easier to use than either CORBA or DCOM, and this may be one of the main reasons that developers have taken to it in such a short space of time. Neither the leading CORBA vendors nor Microsoft are actively promoting RMI since it competes with each party's favoured solution therefore both have tried to discourage its growth. Under pressure from the CORBA world, Sun announced its willingness to somehow merge RMI with CORBA's IIOP [OMG97] but as RMI is designed solely for Java while IIOP is more generic, there are technical issues that make this less than a perfect solution. Microsoft simply refused to support RMI. Given that Microsoft provides the operating system for the vast majority of desktops in the world today, this poses a bit of a problem for organizations that wish to use RMI but one solution for companies that choose Java is to use Netscape's browser, which includes RMI support, rather than Microsoft's Internet Explorer, which does not. For organisations standardising on the Microsoft browser, using RMI means installing the necessary library on every client as it is not built in. RMI lacks security among other important services which perhaps CORBA supports but if these limitations are acceptable, and one has a means to guarantee the availability of the RMI libraries on both clients and servers, and, most important, if the objects on both sides will be written in Java, then RMI is a good choice. The middleware discussed in this section are based on the RPC model. RPC technology as the name suggests, amounts to simply calling procedures that live elsewhere. It is an easy and popular paradigm for implementing the client-server model of distributed computing. In general, a request is sent to a remote system to execute a designated procedure, using arguments supplied, and the result returned to the caller. Parameters can be established for the remote procedure and values can be returned from it. This provides a sort of poor man's application partitioning, letting you place parts of your application on other hosts or processors. A chief disadvantage of using RPC-based middleware protocol is that most RPCs execute synchronously therefore a call to an RPC will usually block application execution until the RPC returns. Vendors have avoided this by executing the call on its own thread, but the underlying technology is still basically synchronous. There are situations where an RPC must finish its task and situations where the RPC does not need to complete therefore applications have to be customised to allow for this, needlessly complicating them.

The main problem with the above conventional middleware systems (i.e. CORBA, RMI and DCOM) is that they have been developed to provide a homogeneous access to remote entities independent of operating systems or hard architectures. Typically, these middleware try to provide as much functionality as possible, which leads to very complex and resource consuming systems, which are not suitable for small devices. Approaches to solve this problem do exist and are highlighted below but it must be remembered that conventional middleware was mainly designed for stable network environments, in which service unavailability is a rare event and can be treated as an error.

1.8.3   Dynamically Reconfigurable Middleware

Extending conventional middleware systems to dynamically reconfigurable middleware systems (e.g. [Blair01], [Blair00], [Becker00] and [Ledoux99]) enables such middleware to adapt its behaviour at runtime to different environments and applications requirements (e.g. how marshalling is done). For instance, changing the communication model or invoking new protocols for incoming and outgoing messages is typically not supported. One exception to this is the Rover toolkit [Joesph97], which provides functionality for a queued RPC (QRPC) concept that is layered on top of different transport protocols alongside other reconfigurable mechanisms. Most existing reconfigurable middleware concentrate on powerful reconfiguration interfaces with little support for smaller resource-constrained mobile devices.

1.8.4   Middleware for Resource-Constrained Devices

The resource restrictions on mobile devices prohibit the application of a full-fledged middleware system. One way to overcome this would be to restrict existing systems and provide only a functional subset (e.g. [OMG02a], [Roman01a] and [Schmidt01]) leading to different programming models or a subset of available interoperability protocols. Another option is to structure the middleware in multiple components, such that unnecessary functionality can be excluded from the middleware dynamically. One such example is the Universally Interoperable Core UIC [Roman01b]. UIC is based on a micro-kernel that can be extended to interact with various middleware solutions but the protocol is determined prior to communication and dynamic reconfiguration is not possible.

1.9   Problem to be addressed by Media Developers

Pervasive computing environments consist of multitudes of heterogeneous devices, both stationary and mobile, with different and dynamic changing capabilities and specific ways to access them. One crucial device capability is the ability to communicate and interact with other devices such as in spontaneous networks with changing members due to the communication range [Becker03]. The network interfaces are highly heterogeneous ranging from infrared communication to wired connections. Interoperability protocols are tailored to specific requirements as well, e.g. a sensor does not need to implement a complex interoperability protocol but can simply emit its data periodically as events. To summarise, devices interact by forming spontaneous networks using different network interfaces and interoperability protocols. Membership in these networks is temporary and network related properties like communication cost and bandwidth change dynamically. Distributed applications in this scenario are structured into application objects, or services, interacting with each other. Services in turn use device capabilities or further services, which are provided by either the local device, or by remote interaction with other devices. From the application's point of view, one of the main challenges is to use services and capabilities with changing availability. In addition, even a service that is both functional and reachable can become unavailable. Take for example a video presentation system integrated into a Bluetooth equipped PDA. If the user leaves the Bluetooth base-stations wired link area, the video stream becomes unavailable, because the user cannot get a signal. Existing middleware platforms typically address portability of applications via standardised interfaces for remote service interaction, e.g. via stub and skeleton objects, and interoperability of applications across different middleware platforms via interoperability protocols.

A common method at present for companies attempting to produce applications for mobile devices is to port PC based applications to wireless devices. The wireless communication environment is characterised by lower bandwidth than the wired Internet, and by sporadic network disconnects. Therefore to create a compelling user experience, new applications need to be developed and optimised for the wireless platform. For the sake of vendor neutrality, scalability, user experience, and time to market, those applications will consist of a rich client deployed directly on the mobile device, and of a server application running at the back-end. The most likely adoption of Java technology for this type of solution will be J2ME[1] (Java-2 Micro Edition) on the mobile

---

[1] http://java.sun.com/j2me/docs/

device, J2EE[1] (Java-2 Enterprise Edition) application servers at the back-end's, and a 2.5 or 3G [3GPP01] network between clients and servers. J2ME only provides rudimentary communications interfaces (e.g., TCP/IP streams, HTTP streams, and datagram connections) for internetworking wireless clients with servers. In order to cope with sporadic network disconnects, with off-line service usage, as well as with the peculiarities of the various wireless networks, developers need to devote a substantial amount of their time to solving technical networking issues, instead of focusing on the business problem to solve. Thus, there seems a need for a mobile middleware platform aimed at enabling reliable, secure, and scalable services in a wireless new-world. Mobile services could be deployed quicker, by taking advantage of the offline operation, data synchronisation, reliability and security mechanisms provided through a mobile middleware framework. This middleware would overcome the more difficult coding against the raw communications APIs provided by generic middleware, thus time to market is reduced and the user experience is improved. Three distinct attributes which can be derived from the core of this body of work are as follows:

Uniform Reduced Instruction Set Programming API: while classical middleware addresses uniform access to remote services the additional heterogeneity of specialised device capabilities requires similar abstractions, e.g. proxy objects, in order to access different device capabilities in a uniform way independent of the underlying platform. It is also the intention of this framework to provide more control over an application's adaptive behaviour during its entire life-cycle. Many different sources of information are available to help decide how a system should adapt but this information is spread out over multiple places, times and people making it difficult to retrieve and usually ultimately ignored. Our architecture intends to organise and disseminate this information between parties in order to enhance the effectiveness of adaptive decisions

Flexible Protocol Support: the service model of a middleware, e.g. remote procedure call or events, is typically reflected in its underlying interoperability protocol, e.g. using request/response messages or emitting event messages. Current devices and systems frequently require the integration of a variety of such service models that are reflected by their correspondent interoperability models. A decoupling of the service model from the interoperability model used by the middleware can help to bridge these interoperability domains. Additionally, this allows different communication paths for the incoming and outgoing messages.

---

[1] http://java.sun.com/j2ee/

31

For example, in the case of two devices communicating via infrared in order to save energy. If the infrared link breaks due to obstacles or distance and a wireless radio link still exists, communication can continue. This can be either achieved by providing one interoperability protocol over different network interfaces or by the abstraction of different interoperability protocols that allows flexible usage of existing technologies.

Tailorable: To be useable on all kinds of devices found in future scenarios, the middleware has to be tailorable to the device at hand, a mobile device as well as a desktop. The core functionality should be small enough to be executed on a mobile micro-device platform, but easily extensible to use the capabilities of resource richer devices. Two additional requirements similar to tailorable are to accommodate traffic heterogeneity and application heterogeneity. To accommodate traffic heterogeneity, the middleware should provide support for non-streaming protocols and streaming protocols at a continuum of rates. To accommodate application heterogeneity, the middleware should not force a particular application style (such as synchronous over asynchronous) but should be flexible to accommodate varying styles. Adaptation on the fly is also feasible on today's hardware. It is more efficient to compute a different representation on demand rather than storing a set of pre-recorded video representations and switching between them during transmission.

The above requirements can be summarised as *Operational Transparency* and *Performance Transparency*. Operational Transparency means that no action needs to be conducted by users due to host movement. This can be achieved by detecting host movement and performing actions that ensure service continuation at the mobile host's new location. Operational Transparency by itself does not make any quality of service guarantees. Performance Transparency ensures that protocol service should continue with a similar performance should a mobile host migrate to a new location. Methods of ensuring that a framework has performance transparency include optimal routing of packets to and from the mobile hosts, efficient and robust migration procedures and efficient use of network resources such as transmission bandwidth and processing. There have been many component-based environments proposed to support the building of architecture-based applications which support dynamic configuration mechanisms [Blair00b, Kon00, Rejaie00, Yuan01] but very few projects have validated the effectiveness of dynamic configuration with resource intensive applications such as multimedia frameworks. The next section will discuss existing approaches to the problem discussed above.

1.10   Existing Approaches to the Problem

The research area of a streaming media middleware framework for mobile clients tends to overlap a multitude of research domains including the network topologies for transporting the media, algorithms to encode/decode and make informed decisions, protocol stacks which carry out those decisions, middleware frameworks to encapsulate the specific classes and mobility policies for seamless interconnection of mobile devices. Next, a variety of solution attempts under the problem areas discussed in section 1.1 are examined.

The IBM BeanExtender [IBM97] system allows customisation of the behaviour of JavaBeans, which do not require access to source code by adopting a metaobject protocol approach to structuring wrappers for JavaBean components [Haefel01]. This approach is based upon the JavaBean event model and component pattern limited to components that follow the JavaBean pattern. Also, while their mechanism allows a form of dynamic customisation, it requires pre-processing of the JavaBean in order to support the addition of behaviours. This does not allow the client to use unprocessed components. OpenJava [Wu96, Tatsubori00] and MetaJava [Golm97a, Golm97b and Golm99] are also reflective approaches. OpenJava uses a pre-processor that performs source-to-source conversions to add reflective behaviour to Java programs and does not make use of the java.lang.reflect introspection interface that was added to the language in JDK1.1. MetaJava provides users with a wide range of reflective facilities for binding metaobjects to objects but like many existing solutions; MetaJava requires the use of a non-standard Java Virtual Machine.

Wireless mobile devices have created a new platform for distributed media processing where mobile users have access to information stores independent of location. The MOWGLI project has implemented a data communication architecture for a pan-European GSM-based mobile data service [Liljeberg96, Kojo97] concentrating on the architectural aspects, which support the mobility of clients allowing them to operate in a weakly connected mode, thus hiding the problems of the wireless connection [Alanko97]. MOWGLI focuses on GSM-based data rather than IP wired-cum wireless networks. Aspects of the MobiWeb [Margaritidis00] architecture are quite similar in many respects to MOWGLI. The issue of allowing mobile hosts to find some point of connectivity when they move around is addressed by the SPINACH project (Secure Public INternet ACcess Handler) [Poger97, Zhao01, Roussopoulos03].   SPINACH provides access control that allows authorized users to connect their laptops to a network thus making it possible to leave network ports or wireless networks available in public areas and know that only desired visitors will be allowed to access the network. This project tends to focus mainly on the security issues of mobile network access. Migrate is an end-to-end framework for Internet mobility developed at MIT [Snoeren01]. Migrate provides a unified framework to support address changes and disconnectivity allowing legacy applications to adapt to highly-mobile environments, and provides mobile-aware applications with a robust set of system primitives for disconnectivity support, conservation of resources, and rapid re-instantiation of network connections. Disconnection is treated as a fundamental component of mobility, applications are enabled to gracefully reduce their resource consumption during periods of disconnection and quickly resume sessions upon reconnection. Mobile agents are investigated in [Thai00, Bellavista01, Valenzuela02 and Bellavista03]. The work of [Kotz00] takes an analytical approach to the performance of data streams arriving at a common gateway which are then forwarded to mobile devices through a wireless channel. Kotz proposed the use of mobile agents originating from the clients device, to migrate from the device to the common gateway. The mobile agents then perform data filtering at the gateway as the streams arrive off the network. Migrate and the above mobile agent research seems to focus on address migration (i.e. mobile hosts migrating to another network location) and agent migration, at the expense of a more complete middleware framework thus limiting users in their application development.

In systems such as described in [Kojo95, Zenel95, Yuan01 and Karrer01], streams are basically decomposed into two connections with an intermediate computer(s) connecting the end systems. When the bandwidth of a wireless network becomes low, an intermediate stationary computer filters data from a high-speed network before sending the data to a mobile computer. The intermediate computer filters the data according to the content such as video data degradation by reducing the size of each video frame. The solution answers the problem caused by drastic changes of network bandwidth, but it does not solve the problem when network bandwidth is changed dynamically during the execution of applications. In addition, these systems do not provide programming supports for creating intermediate filters systematically. Numerous other researchers have proposed adaptive mobile protocol mechanisms but little implementation of such mechanisms exist.

The approaches taken by middleware solutions, such as WAP [WAP02] or MASE [Kreller98, Meggers98] is to encapsulate the weaker process by creating simple APIs or reduced operating system calls. In this approach, applications are forced to consider processor limitations by using a correspondingly limited set of functions. MASE is a Middleware layer intended to enable multi-media applications on mobile networks with focus on QoS handling over cellular phone networks for mobile devices. This approach require rewriting of client code, which can be a drawback given the large existing base of applications. Limitations in displays are the most commonly addressed issue on the client side. Again, the most common approach for gateway solutions is to handle the problem in the proxy. Filtering and compression are the methods of choice. MASE, for example, uses filtering to remove unsupported content types from the data stream going to the client [Kreller98]. Filters must be configured for the mobile client specifying which data types and what resolutions can be accepted.

The Rover toolkit [Joesph97] provides relocatable dynamic objects and queued remote procedure calls for building mobile applications. The focus of the toolkit is to build mobile applications that support disconnected operation. The relocatable dynamic objects allow dynamic configuration of data filters that enables mobile applications to be adapted to drastic changes of configurations of mobile computers. Methods associated with an object are invoked on the objects data using Tcl scripts. Tcl scripts run an order of a magnitude slower than for example C counterparts. There have been numerous examples of proxies acting as intermediaries between clients and servers for HTTP [RFC2068-97, Kassler00a, Kassler00b]. Proxies can also be used to hide the effects of error-prone and low-bandwidth wireless links [Fox96, Liljberg96, Zenal97, Huang02 and Yoshimura02] applies the proxy mechanism to mobile environments where filters on an intermediary, delay and transform data between mobile and fixed devices. The filters are part of the application, complicating their reuse and making it difficult to support legacy applications. The InfoPad project [Infopad99] used an extreme approach with proxies where all intelligence located in the infrastructure and PDAs are used basically as dumb terminals. The Application Level Gateway (ALG) [Long96, Amir97] is an attempt to handle the disparity that exists between end-to-end systems, and the networks that connect them. The ALG uses dynamic and transparent bandwidth adaptation of the continuous video bit-stream through explicit mechanisms such as video transcoding and rate control. The video gateway transcodes intra-H.261, Motion-JPEG (MJPEG) and NetVideo (NV) and the audio gateway transcodes pulse coded modulated (PCM) audio. The application level gateway being a static entity requires each node in a network hosting video gateway software to be individually installed and configured. Other proxy-base projects include VGW [Amir95, Amir01] which can transcode RTP video streams from high bit-rate MJPEG format into 128 Kbps H.261 video streams which are more suitable for MBone sessions. Hemy et al. [Johanson01] developed an RTP to HTTP gateway which interconnects a multicast network with the WWW and enables web clients to receive video streams from multicast conference sessions. In [Hemy99], an MPEG specific proxy-based content adaptation assigns priority to I-frames over P and B frames by selecting frames to drop when congestion occurs. All of these projects were proposed to solve a particular problem and focussed on a specific encoding format or protocol conversion. The objective of the framework presented here is to provide proxies with the additional flexibility required to implement any of these adaptations. Moreover, such flexibility aims at providing content based adaptation of video streams in order to take into account both QoS problems and user requirements.

Multicasting [RFC3376, Deering89, Deering90, Kunz02, Phan02] has received considerable attention due to the interest in collaboration technologies but; much of the work revolves around either low-level communication protocols or groupware applications, which typically simulate multicast through a series of unicasts [Rejaie99, Handley00]. Protocols that build on multicast specially designed for continuous media transmission include ST-II [Mitzel94] and RSVP [Crawley98, Terzis00]. These protocols support multicasting and resource negotiation appropriate for continuous media, but no specific implementation mechanisms (e.g. for multicast route set-up or resource reservations) exist thus forcing developers to implement these mechanisms outside the protocol. Group based multicast systems such as described in [Maffeis96, Maffeis97, Tai02] attempted to address the problems that face mobile devices that are intermittently attached to a network by building a hierarchy as receivers join the multicast group. Servers are connected to other servers higher in the hierarchy, and eventually back to the source with delivery from servers to children through unicast reliable protocols. One of the systems - GTS facilitates this flexibility by specifying communication end-points as URLs, including a "ticket" indicating the multicast channel. When disconnected hosts are unavailable, the server must spool the message until it can be delivered thus GTS is not prompt in its delivery, but robust. These systems therefore are more useful for replicated databases or software distribution rather than streaming media. Scalability issues also arise, as senders must contact the single sequencer server directly reducing the number of senders that can be supported.

The QoS-A protocol distributes multicast data through carefully selected nodes in a hierarchy, which are equipped to filter multimedia streams to reduce their demands on receiver [Campbell98, He02]. An MPEG-2 stream can be reduced to an MPEG-1 stream, or MPEG-1 could be reduced to contain only I-frames. To facilitate feedback, the protocol allows both clients and filters to notify upstream filters of their limitations. If all the clients of a filtering node have a common limitation, for example, all can display only monochrome video, the limitation is passed up the hierarchy towards the source. Filters reduce the bandwidth of a stream until it meets the limitations of a destination. End-users are unable to explicitly state their capabilities such as in the case of a client capable of only audio to upstream filters. These multicast based solutions only address network bandwidth; applications do not consider clients' hardware limitations and software incompatibilities. Moreover, they require modification of the software installed on Internet hosts [Kassler01a, Kassler01b].

Hierarchical encoding has been studied extensively in the area of signal processing and as a potentially effective mechanism for congestion control for high-speed networks carrying digital continuous media [Kompella93, Delgrossi95, Cheung96, Heinzelman00]. Receiver-based layered transmission allows the delivery of varying scaled levels of video over multicast to accommodate heterogeneity while performing coarse-grain congestion control [Wu97, Mertz97]. This method does not allow for fine-grain delivery control and the ability for additional intelligence to be 'injected' into the data path to increase the transmission rate. [Feng99] proposes an adaptive smoothing mechanism which combines bandwidth smoothing with rate adaptation. The transmission rate is shaped by dropping low-priority frames based on prior knowledge of the video stream which is meant to limit quality degradation caused by dropped frames but the quality variation cannot be predicted as is the case with many other approaches in this area..

Supporting Variable Bit Rate (VBR) video along with audio and data over bandwidth-constraint networks continues to be a formidable problem [Handley00]. The difficulty arises because VBR video is unpredictably bursty and because it requires performance guarantees from the network. Resource Reservation schemes may work well for Constant Bit Rate (CBR) traffic but there is no agreement on which strategy should be used for VBR traffic. Since VBR traffic can be delay sensitive, a resource reservation scheme seems to be the correct choice, but because video is bursty, if resources are reserved according to peak rates, the network may be under-utilised if the peak to average rates ratios are high [Bahl97, Fankhauser99]. Several bandwidth partitioning strategies that allocate bandwidth fairly for different traffic classes while attempting to achieve maximum network throughput have been proposed [Schwartz95, Tan01]. Previous studies of these techniques in wireless networks have focused on the co-existence of data and voice traffic, while packet video has generally been ignored [Bahl97, Wu02].

There are numerous commercial multimedia platforms on the market today that provide an end-to-end solution for streaming video over wireless networks. One of the leaders is Emblaze Systems whose Emblaze technology enables encoding and playback of live and on-demand video messages and content on PCs, PDAs, video mobile-phones and interactive TV [Emblaze03]. The Emblaze Wireless Media Platform comprises an Emblaze Server and Emblaze Encoder. Emblaze Encoders can be geographically distributed and deployed where content is created. Another market leader is Packet Video which provides MPEG-4 standards-based video and audio transmission over bandwidth-limited, error-prone wireless networks. Packet Video's PVAuthor 2.0 software supports input of popular digital content file formats [PV03]. These systems are built specifically to deliver video to handheld devices and indeed produce some impressive results. Chameleon differs from these systems in that it focuses more on the issues of the interplay between a mobile service platform and a multimedia delivery platform.

A range of middleware technologies exist, including the Common Object Request Broker Architecture (CORBA) [Hector02, Vinoski97, Fingar00], Globus [Foster01], and DCOM [Knudsen00]. CORBA enables objects to interact in a language and platform independent manner through an Interface Definition Language (IDL). An Object Request Broker (ORB) allows clients to issue requests on an object where the ORB locates the object, transmits the request, prepares the object implementation for receiving and processing the request, and returns results to the client. A problem with CORBA is that the architecture adopts a traditional black box approach such that the platform implementation is hidden from the application [Blair98]. This can prove to be a major problem in development of distributed systems as it is essential to have (selective) access to the underlying implementation [Chaung97, Garcia99]. It has been proposed to use CORBA middleware on mobile devices and there is a low-footprint CORBA implementation being worked at presently titled 'Wireless Access and Terminal Mobility in CORBA' [Raatikainen03]. The CORBA specification was first presented back in 1992, thus CORBA was not designed for mobile communications and is used almost exclusively in server environments over corporate networks [Krishna03]. It can be argued that the standard gained popularity until about 1998. With the success of the Java platform, attention has been moving away from CORBA, towards middleware solutions which are better integrated into the Java platform and are less complex than CORBA (e.g. RMI and Enterprise JavaBeans) [Grant02]. CORBA was initially conceived for computing environments in which multiple programming languages are used but in homogeneous environments, such as J2EE or Microsoft .NET, developers tend to prefer the middleware tools integrated into that environment.

In addition, researchers at CNET have developed an extended CORBA platform to support multimedia through the concept of recursive bindings [Blair98, Blair02], which makes mobile applications adaptive by changing policies for caching, prefetching and reducing of data according to machine configurations. CNET does not modify the structure of mobile applications when the execution environment is changed. GLOBE is a middleware framework that allows objects to be physically distributed and to consist of fragments at several nodes so that distribution and communication between fragments are hidden from other objects [VanSteen98, VanSteen99]. The distributed objects provide scalability to wide-area distributed applications such as replication and caching for web documents and services exist for locating and downloading fragments, through which objects are accessed. The API makes it difficult to integrate complex transmission mechanisms such as feedback or compression into GLOBE. These proposed scenarios are similar to this work but their implementation relies on the CORBA platform and they do not report any performance evaluation. There has being considerable research into extensible and adaptable operating systems such as Spin [Fiuczynski98], Exokernel/Aegis [Engler98] and U-Net [Oppenheimer98]. The aim being to introduce flexibility into operating system structures to allow, for example, the addition of new services. In general, this research has not considered the requirements of mobile multimedia applications. It is preferable in many situations to implement adaptation in the middleware so as to attain a platform independent means of achieving adaptability.

Other related work is within the area of nomadic computing such as the Oxygen project [Oxygen02] from MIT which targets in the means of turning a dormant environment into an empowered one that allows the users to shift much of the burden of their tasks to the environment. The project mainly focuses enabling technologies such as new embedded distributed computing devices, speech access technology, intelligent knowledge access technology, collaboration software, automation technology for everyday tasks, and adaptation methods. [Dertouzos99]. The Future Computing Environments (FCE) Group at Georgia Tech [FCE03] is attempting to break away from the traditional desktop interaction paradigm and move computational power into the environment that surrounds the user. They differ in that they place a larger emphasis on knowledge about the user and the environment that surrounds the user [Dey01]. The Portolan project in the University of Washington at Seattle [Portolano02] focuses on Context Aware Computing, which attempts to coalesce knowledge of the user's task, emotions, location, and attention, which they identify as an important aspect of user interfaces [Esler99]. The 2K and Gaia from the University of Illinois at Urbana-Champaign [2K03] manages and allocates distributed resources in order to support a user in a distributed environment. Their objective similar to the above projects is to achieve a better fulfilment of user and application requirements. The architecture encompasses a framework for architectural-awareness so that the architectural features and behaviour of a technology are reified and encapsulated within software [Roman00].

The Mobile Computing Group at Stanford University (MosquitoNet) [Mosquito03] has developed the Mobile People Architecture (MPA) that addresses the challenge of finding people and communicating with them personally, as opposed to communicating merely with their possibly inaccessible machines. The main goal of the MPA is to put the person, not the device that the person uses, at the endpoints of a communication session [Maniatis99]. MPA also focuses more on Mobile IP implementations rather than dynamic reconfiguration. Finally, the Monads project at the Department of Computer Science in the University of Helsinki [Monads03] has addressed wireless communication on different protocol levels, adaptation to changes in the communication capabilities, and short-term (1-30 min) predictions of available communication capabilities. The Monads approach is different from Chameleon in that Monads recognises the shortcomings of reactive adaptivity by using predictions in order to be proactive. In order words, the Monads system tries to carry out adaptive actions before the connectivity goes down or to postpone the actions until the connectivity improves [Raatikainen99, Campadello00].

1.10.1   Summary of Weaknesses in Approaches

Problems with RSVP and other end-to-end per-flow resource reservation techniques has been well documented in the literature with regards requiring core routers to maintain individual flow states leading to scaling problems (not to mention actual deployment). In addition, existing middleware and multicast protocol approaches tend to concentrate on the delivery of efficient narrowly defined solutions at the expense of generic adaptable frameworks which can be applied in wider ranging domains to cope with existing and unforeseen occurrences. Hierarchical methods do not allow for fine-grain delivery control and the ability for additional intelligence to be 'injected' into the data path to increase the transmission rate. Reflective approaches require the use of non-standard Java virtual machines which again prevents the ready deployment of applications (which could make use of existing standard java virtual machines in routers, web browsers, JDK's etc). Proxy solutions are often written in slow scripting languages (hampering performance) or the filters are part of the application complicating their reuse and making it difficult to support legacy applications. Mobile frameworks seem to focus on issues such as mobile hosts migrating to another network location, at the expense of a more complete middleware framework thus limiting users in their application development.

Good bandwidth allocation schemes rely on dynamic allocation of bandwidth to achieve high utilisation. While dynamic allocation at the burst level provides good statistical multiplexing, it performs poorly for connections that require a certain quality of service. Factors against using CORBA on mobile devices include the fact that it is inherently based upon a *request/response* synchronous communication model which leads to a misfit between this communication model and the nature of wireless networks. Wireless networks are packet based with frequent network disconnects where transmission speeds and delays vary a lot. The CORBA communication model assumes that communication links are stable, have low error rates, and that network disconnections occur only rarely. This is an acceptable assumption in fixed wired environments, but not in wireless networks. CORBA is also based on a classical client/server interaction model. Wireless services often require push notifications, queued communication, and peer-to-peer sessions. It can be argued that these services can be granted through the CORBA Notifications service and Time-Independent Invocations (TII) but these mechanisms are too heavy weight and complex to be deployed on mobile devices [Pyarali02].

1.11   A new Approach to the Problem

From literature surveys it has become evident that the problem of resolving heterogeneity in mobile multimedia devices is not a trivial one as evidenced by the problems discussed in section 1.1 and the solution attempts discussed in section 1.1. This section describes the solutions to the problems discussed earlier.

1.11.1   A New Approach Overview

The framework presented here is an extension of the common client-server architecture where one host (server) provides certain services or applications of which another host or device (client) makes use. The proxy (gateway or mediator) is another host logically placed between the server and client in order to mediate, route, or otherwise facilitate communication between them. The client-proxy-server approach can be implemented at different levels in the protocol stack but when it is applied to the network or transport layers the result is a mobility-transparent solution. Mobility solutions that are transparent to the application are desirable because of the reduced impact to the applications in that they allow most of the fixed network (Internet) to remain architecturally unchanged. They also allow traditional protocols and software application designs to operate as they always have. They support the use of new and optimised protocols between the fixed network and mobile clients. Given the fact that fixed servers have thus far been several orders of magnitude more powerful than mobile devices, these solutions also allow the use of such powerful servers to perform resource intensive filtering, compression and protocol conversion tasks. This framework generalises the data stream adaptation approach where multiple adaptive proxies filter the stream along the path. Each adaptive proxy further decomposed into an Event Manager (EM), a System Monitor (SM), and many more components which attempt to transform the data stream and make requests for additional resources as required. The SM is also informed by the EM of changes in resource utilisation and network conditions. Proxies are placed at the edge between the traditional fixed network and the mobile network in order to modify and adapt content to suit each mobile device. The greater computational and system resources of the gateway machine is used to "digest" the data stream and pass on a version which is suited to the wireless bandwidth and end-user device capabilities.

While transparent support for wireless networks and mobile clients is favoured as a non-intrusive method for integrating mobile networks with fixed networks, it can also be argued that the mobility cannot be fully exploited without the application's involvement. Therefore, the load created for proxies carrying out the transparent adaptation may create a scalability problem as wireless devices increase in number. The proxy model can result in large amounts of data being passed over networks, followed by resource intensive computations on the proxies, which can ultimately result in data being discarded or modified. This suggests that end-to-end solutions that involve the application in the adaptation decisions may provide better scalability [Joshi00]. Infrastructure solutions that are meant to be transparent can also be designed to address scalability. This leads to the concept of applications that are aware of the mobile environment they are operating in. The solution framework presented here achieves this by structuring mobility support into a middleware layer that the application can use to be notified of, and respond to, network changes. The use of a middleware approach will provide greater flexibility for application development alongside the useful functions that proxies provide as previously mentioned. Applications can be developed for many purposes and as long as they call on the functions of the middleware they can support mobility.

1.11.2 Detailed Approach Overview

Multimedia has varying optimal transport methods; but the traditional methods employed by transport protocols are to ship all data through identical protocol stacks. A paradigm is proposed whereby media are transported through an optimised stack constructed solely for that media/medium allowing improved multimedia Quality of Services to be achieved even at run-time. In our approach, for making applications adaptive, the applications are composed from small (micro) stack objects and the composition is reconfigured when the operational environment is changed. Each stack object has a function such as filtering and caching and a protocol profile manager can dynamically reconfigure the object compositions in an effort to achieve increased performance. Configurable end-system protocols support a wide range of application requirements and increase protocol performance by decreasing protocol complexity. The dynamic configured protocols will represent light-weight protocols, as all unnecessary protocol functions will not be present. This approach provides a systematic way for constructing adaptive mobile applications.

Systems exist which support interactive applications dealing with synchronised time-based media, but most existing systems operate in a stand-alone environment. Inconsistent distributed multimedia services, which require sophisticated levels of interfacing in order to reflect different aspects, add to the problem. A means to alleviate this problem is to offer an encompassing framework in which all services would be offered under a unifying paradigm. Often, it may be desirable to choose from a variety of mechanisms when the application is running, such as the case of a mobile device moving from a LAN into a cellular network. As a result, the available bandwidth and the error rate change significantly. Ideally, the middleware should adopt bandwidth-conserving mechanisms, such as compressing the requests, i.e. trading off CPU for bandwidth in order to adapt itself to environmental changes. Fluctuations in throughput and delay may be experienced due to congestion even while connected to a particular network, (e.g. as witnessed in the Internet) therefore it is important that systems can adapt to the offered network QoS. Adaptation can take place at a range of levels in a system and this work is concerned with support for adaptation in the middleware platform, i.e. the software layer above the operating system offering a platform independent programming model thus hiding problems of heterogeneity. In particular the design of a middleware platform for streaming media is considered, which allows the application to inspect the current level of QoS at various points, and enables applications to dynamically adapt the behaviour of the underlying platform in response to QoS changes. Adaptable applications generally degrade the precision of the media in order to cope within lower bandwidth environments, but the timing constrains of the media should be left intact thus ensuring the timeliness of the application. Acceptable mobile adaptation methods should ensure the delivery of the expected communication quality without wasting valuable radio spectrum resources and with practical distribution of the processing load.

Proxies located at the home agent of each mobile client can also offload processing for memory-constrained devices. The home agent is responsible for activities such as message queuing and forwarding, access control to streaming sessions, message encryption and protocol translation among others. In order to allow the smallest possible client footprint, the home agent (and proxies) relieves the client library of most of the work of maintaining state information about active sessions, filters and general system monitoring facilities. Micro protocol stacks have been adopted to simplify the interoperability process, by allowing code for dynamic stacks to be written once and placed on repositories where they can be shared. This allows thin clients such as palm tops, WMP phones etc. to download new protocol stacks. A flexible protocol system allows the dynamic selection, configuration and reconfiguration of protocol modules to dynamically shape the functionality of a protocol in order to satisfy application requirements or adapt to changing service properties of the underlying network. Flexible end-to-end protocols are configured to include only the necessary functionality required to satisfy an application QoS requirements for the particular connection. Another benefit is that the actual code footprint for all the stack elements can be minimised with only the required stacks being actually in place on the client machine (rather than the entire potential library) thus allowing protocol stacks to fit on limited memory devices that might otherwise not have been possible. These mobile proxies have the task of providing optimal service to heterogeneous clients through reconfigurable optimised protocol stacks. A proxy is an application that is placed at the far end of a low bandwidth connection, to downgrade a high bit rate stream, so that it can fit through the connection. Proxies (or Transcoders) for video and audio can reduce the frame-rate and image quality, and audio transcoders can re-encode the audio signal using a higher compression bandwidth scheme. Apart from changing the bandwidth requirements of a stream, a transcoder can also introduce or remove forward error correction to counter packet loss. In scenarios such as multicast broadcasts, transcoders are positioned at the problematic cellular network link's base station, to re-encode the stream to use lower bandwidth. The resulting stream is then re-multicast to a new address. If all receivers beyond that link tune into the new customised stream, then there will be no bandwidth wasted, as the original stream no longer has to traverse the problematic link. Mobile IP solves the problem of mobile clients moving between different wireless LANS throughout the communication process. The mobile hosts use the proxies as a gateway to the Server. Proxies allow clients and servers to remain unchanged and eradicate or minimise the disruption to either when network environmental conditions change. A proxy appears as a server to a client and as a client to a server allows easier deployment than solutions requiring modification of numerous clients or the server. Proxies can be applied to many performance discontinuities on the edge of networks. This proposal invokes knowledge at both the client and the server to implement dynamic proxy based systems for content distillation and protocol enhancement.

Various approaches to providing a solution to the burstiness of VBR video traffic generally lie somewhere between traffic classes sharing the entire bandwidth and bandwidth being divided into distinct portions with each portion corresponding to a particular traffic class. All classes sharing the bandwidth means that a temporary overload in one traffic class can degrade the quality of other classes whilst dividing bandwidth into portions can be wasteful if the predicted demand for a particular class is greater than the actual bandwidth. Chameleon offers a hybrid approach where bandwidth is dynamically allocated to match the varying traffic load so that prioritised media streams are able to 'borrow' bandwidth from lower priority streams. It has been shown that such hybrid schemes provide optimal performance over a range of offered loads both in micro and macro-cellular environments [Kim93, Badrinath00, Joshi00, Matthur03].

A heterogeneous mobile client population can be catered for by providing various qualities of data and sending these media to separate multicast groups so that clients can 'pick and choose' a Quality of Service (QoS) in accordance with available resources. This mechanism offers movement between multiple multicast media groups and transcoding of media within groups (no need to join a separate group). This book presents two innovative QoS mechanisms which are named PQT and SQT. A Primary Quality Transformation (PQT) technique, can assume responsibility for coarse grain adaptation decisions by moving between multicast groups upon violation of group bandwidth limits. Secondary Quality Transformation (SQT) complements PQT by assuming responsibility for responding to quality fluctuations within each group. Both techniques can be optimised to work with priorities being assigned to differing streams within a link in order to allow different streams to be rate controlled according to application-implied importance. Mobile transcoding proxies also tackle the problem of heterogeneous group communication by being dynamically loaded and activated within the network to provide for individual services within a heterogeneous multicast group. Innovative steps are the design of adaptable protocols tailored to a particular media stream within each mobile intermediate system that allows for efficient and flexible service transformations. Mobile proxies specialise the interface of servers for mobile hosts since the interface of servers is difficult to be changed, and the interface is designed for wired networks. Although application specific protocols within the mobile field have been advocated [Heinzelman00], to our knowledge, no framework exists for the provision of environment specific downloadable location based protocol stacks to mobile devices.

As stated earlier, true Quality of Service at present is not in place over the Internet, as it requires end-to-end support but there is a second kind of QOS where bandwidth is not reserved, but is given priority over other traffic. This is known as differentiated service [Trimintzios01]. A run-time quality of service renegotiation technique can be designed to give specific media streams higher quality when overall bandwidth availability diminishes. One approach is to adapt the video encoding to a suitable format at run-time. This can be achieved with a client-proxy-server architecture and dynamic protocol stack deployment mechanisms. Another supporting mechanism is priorities. This is a simple form of traffic differentiation and can be introduced into the network without incurring a standardization process, simply by injecting the appropriate service logic into the media path. During congestion periods media streams are queued as high, normal, medium, or low. Using Priority Queuing, all high-priority traffic is serviced first, then normal, and so on (e.g. Audio, then Video etc). In addition, resource reservation protocols, pricing schemes among others are just some of the capabilities that may be required by future clients. To cater for this, future clients can be provided with the necessary interfaces to allow components to be 'slotted in' at run-time.

The desire here is to create a framework that can evolve and adapt to changing environments similar to living things in their struggle for survival. This can be achieved through a layered Java framework providing an API for a set of middleware services that it abstracts building upon multimedia, mobility, middleware and reflective frameworks to provide a union of these service API's in a streamlined API. The portability of Java make it an ideal middleware language for future systems including 2.5/3G mobile devices and embedded Java systems overcoming the non-portable limitations of the previously discussed protocol frameworks. A simplified programming model (reduced instruction set) with a high level control API allowing elements to be optimised in software with typical client side implementations of <100KB is highly desirable.

1.12   Organisation of Book

The main motivation for the work reported in this book has been described and problems that middleware address have been pointed to. The driving force behind this work has been established. The key problem is that of satisfying the individual QoS requirements of heterogeneous receivers in wired-cum-mobile environments. Please note that the Chameleon framework discussed here is available for free to third-parties for non-commercial use. Please email the author at kj.curran@ulster.ac.uk for further details.

The rest of this book is organised as follows. Chapter 2 introduces the technologies and challenges involved in transporting media over IP networks to mobile devices. Chapter 3 provides a more indepth rationale supporting this work while chapter 4 presents an overview of the components within the middleware framework solution and introduces active service proxies along with dynamic adaptation. Chapter 5 provides a summary.

2   Technologies used in Middleware Solutions

Streamed applications are basically one-way flows of media such as on-demand video and audio services with quality of service (QoS) requirements that must be considered on an end-to-end basis [Zhao01] thus host and network capabilities are equally important in delivering the QoS support required at the application layer.   For media requiring timely guarantees, concerns to be addressed include temporal properties such as delay, jitter, bandwidth, synchronisation and reliability properties such as error-free delivery, ordered delivery and fairness.

Audio quality is highly sensitive to jitter, and video is sensitive to available bandwidth [Apteker95]. For lip synchronisation, audio and video streams need to be synchronised to within 80-100ms for skew to be imperceptible [Jardetzky95]. Packets are effectively passed automatically through to the presentation device. Interpretation of the delivered information is left to human perception; because humans are far more tolerant than computers, lost packets are likely to be perceived merely as a temporary quality reduction [Parr-Curran98c]. Nevertheless packet loss is still a significant problem for isochronous interactions. For example, since a typical packet size is generally above the threshold for audible loss (~20ms), the loss of a single audio packet can be noticeable to the receiver. Resource reservation protocols (see section 1.4.2) are an attempt to resolve these difficulties by allocating resources prior to communication.   Videoconferencing differs from one-way audio and video on-demand in that the communication is bi-directional, and end-to-end delays must be very low (<200 ms) for interactive communication. Standards can now be considered mature having being defined by the International Telecommunication Union (ITU) as H.3xx standards [ITU245, ITU323, and Schaphorst96], and from the Internet Engineering Task Force (IETF) [Willebeek96, Turletti96, and McCanne95]. Each of these standards utilise similar audio-video compression standards but differ in networking protocol specifications.

Audio and video streaming systems currently in the market today such as StreamWorks[1] Vosaic[2] and RealAudio[3]. These tend to be based on proprietary client-server systems [Venditto96] rather than open standards. StreamWorks and Vosaic are designed to work over higher bandwidth LAN connections and not at low modem speeds (e.g. 9600-28,000bps) thus at speeds such as 14,000 baud, these systems have the tendency to degrade to slide-show-type video. Uncompressed multimedia

---

[1] http://www.vdo.ne
[2] http://www.vosaic.com
[3] http://www.realaudio.com

data require a lot of storage capacity and very high bandwidth [Steinmetz95]. Uncompressed audio streams of CD quality is sampled at a rate of 44.1 kHz and is quantized with 16 bits per sample in 2 channels, hence the bandwidth requirements is 44100 x 16 x 2 = 1.41 Mbps. Thus the use of multimedia compression is therefore very essential. Since the source should encode the streams and the destination should decode them, multimedia compression imposes substantial loads on processing resources, such as CPU power [Colouris01].

2.1   Distributed Objects

An Object is a programming abstraction typically considered to encapsulate data and behaviour. Distributed Objects may be accessed over networks and may provide a set of related capabilities where the location of the object is not critical often referred to as services. Distributed objects can allow business processes to be modified or re-implemented without altering applications that use them. They differ from the basic idea of distributed processing in the sense that objects are an enabling technology for distributed systems just as they have been for client/server, multimedia, document processing and other applications. There are several concerns in the case of distributed systems: naming, address-space conversion, transport protocols, interface description, security, latency and quality of service among others [Warfield02]. Each of these issues is complex and objects are the software technology to abstract and deal with complexity. Not only do objects provide a tractable way of organising the complexity of a modern operating system, they can also simplify distributed processing. Objects with their natural combination of data and behaviour and strict separation of interface from implementation make a neat useful package for distributing data and processes to end-user applications. In a Distributed Object Computing environment, application development and management are simplified because the clients do not need to know how the objects are implemented, what languages they are written in, what hardware or operating system platforms they run on, and so forth. Client requests do not alter the code or the intrinsic operation of objects, nor do they establish static relationships with them. The fundamental idea behind interoperable objects is to cross existing boundaries between operating systems, address spaces, machines and languages. The first boundary is the address space. An interprocess object model enables a process in one address space to request the services of an object in another or two processes to share an object in a third address space.

The next boundary is the machine. This is conceptually similar to the address space boundary, but this crossing spans a great deal of complexity. An interprocess object model must be able to translate the data associated with requests between memory models in different address spaces. A technology that crosses the machine boundary must also locate the server object, establish communication with it, set up the request and send it off, then wait for results and return them to the application. These are only the most basic requirements. Then comes the requirement for security, versioning, name resolution and a host of other details inherent in distributing objects across a network. Only object technologies that cross the machine boundary can be truly called distributed object technologies. The ultimate goal is a model that allows objects written in any language to be shared among applications written in any other language, running on any machine in a network, and under any operating system [Betz94, Defago03].

Distributed object component computing represents a computing paradigm in which all information, data and software applications reside on local/remote hosts and are accessed on demand by users [Griwodz99, Joshi00]. This approach to computing offers the potential for users to access 'everything' from 'anywhere', removing the constraints of localised storage of information and applications, and of proprietary hardware and software architectures, provided the necessary standards can be established. In the ultimate version of distributed object component computing, users accessing the network from a device with a minimum set of capabilities would potentially have access to unlimited resources of information and applications, provided they had a connection to the network of adequate performance, in terms of bandwidth, latency and reliability. The applications they could run would not be constrained by the access device characteristics; servers within the network would flexibly partition tasks between the user's access device and machines elsewhere on the network [Badrinath00, Matthur03].

2.2   Video Compression Standards

Common video codec standards for streaming video include H.261, H.263, MJPEG, MPEG-1, MPEG-2, MPEG-4 and MPEG-7 [Richardson02]. In contrast to video codec's for high bandwidth channels such as TV broadcast, Internet codec's require greater scalability, lower computational complexity, greater resiliency to network losses, and lower encode/decode latency. Research continues into scalable, flexible codec's and methods of scaling existing codec's using transcoding and filters and new algorithms specifically targeted at Internet video are being developed [Hemy99].

H.261 [H261] was created for ISDN teleconferencing - in particular for face-to-face videophone applications and for videoconferencing [Khansari94, Richardson02]. The actual encoding algorithm is similar to MPEG but H.261 needs substantially less CPU power for real-time encoding than MPEG. H.261 optimises bandwidth usage by trading picture quality against motion, so that movies with more motion possess lower quality than more static movies.

H.263 is similar to H.261 with improved performance and error recovery. H.263 is designed for low bit rate communication [Streit95, ITUH263-97, Vranyecz98]. H.263 uses half pixel precision whereas H.261 used full pixel precision and a loop filter for motion compensation. The hierarchical structure of the data stream allows the codec to be configured for a lower data rate or better error recovery through advance frame prediction algorithms similar to MPEG called P-B frames.

JPEG [Wallace91, Austerberry02] is designed for compressing either full-colour or grey-scale images of natural, real world such as photographs and naturalistic artwork. JPEG is a lossy compression algorithm that uses DCT-based encoding typically achieving 10:1 to 20:1 compression without visible loss. 30:1 to 50:1 compression is possible with small to moderate defects, while for very-low-quality purposes such as previews or archive indexes, 100:1 compression is quite feasible. In JPEG the entire picture is evaluated and redundant (i.e. similar and interchangeable) pixels are grouped in blocks. The higher the degrees of compression, the more pixels are grouped in a block. At a high level of compression the creation of blocks has a negative effect on picture quality, with the blocks (artefacts) becoming clearly visible in the picture. JPEG compression always entails a loss of data; hence it is impossible to restore the picture to its original condition. In the full view one hardly notices any change at first glance but in a close-up detail view, the artefacts are clear to see. Therefore, the most important thing to consider when selecting the degree of compression is the size in which the picture is to be viewed or printed. The JPEG compression technique is suitable for single pictures but not for video sequences.

MJPEG is an adaptation of the JPEG compression method for video sequences. Data compression is essential particularly for video recordings. Without compression, a 90-minute film with 25 pictures per second would require approximately 120 GB of memory. In M-JPEG (Motion-JPEG) [Stefanov00], each individual picture of the video sequence is compressed in accordance with the JPEG technique. This is particularly good for exact imaging because each single picture is present in full but the amount of memory required is still relatively high.

MPEG (*Motion Picture Experts Group*) achieves high compression rate by storing only the changes from one frame to another, instead of each entire frame [Berkeley02]. MPEG uses a type of *lossy compression,* since some data is removed. But the diminishment of data is generally imperceptible to the human eye. There are four major MPEG standards: MPEG-1, MPEG-2, MPEG-4 and MPEG-7 [Balasub03].

MPEG-1 [Gopalak99, Li98] works over 1-1.5 Mbps offering VHS quality CIF video at 30 frames per second. MPEG requires hardware for real-time encoding but decoding can be done in software at the cost of consuming a large proportion of the processing power. MPEG-1 does not provide resolution scalability and is highly susceptible to packet losses, owing to dependencies in the P (predicted) and B (bi-directionally predicted) frames making it less suitable for video. The most common implementations of the MPEG-1 standard provide a video resolution of 352-by-240 at 30 frames per second (fps) which is slightly below the quality of conventional VCR videos.

MPEG-2 [Senda99, Hemy99] includes support for higher resolution video, scalability and increased audio capabilities over 4-15Mbits/s, providing broadcast quality full-screen video. MPEG-2 requires more expensive hardware to encode and decode than MPEG-1 and is prone to poor streaming quality in the presence of losses, for the same reasons as MPEG-1. MPEG-2 is ideal for high-quality TV broadcast type applications allowing the encoding of up to 8 digital satellite broadcast channels using the same bandwidth as used by one single analog channel. MPEG-2 offers resolutions of 720x480 and 1280x720 at 60 fps, with full CD-quality audio. This is sufficient for all the major TV standards, including NTSC, and even HDTV. MPEG-2 is also used by DVD-ROMs.

MPEG-4 [MPIF03] is a graphics and video compression algorithm standard that is based on MPEG-1 and MPEG-2 and Apple QuickTime technology. Wavelet-based MPEG-4 files are smaller than JPEG or QuickTime files, so they are designed to transmit video and images over a narrower bandwidth and can mix video with text, graphics and 2-D and 3-D animation layers. MPEG-4 was standardized in October 1998 in the ISO/IEC document 14496 [ISO14496]. MPEG-4 defines the coded representation of objects such as: text and graphics; talking synthetic heads and associated text used to synthesize the speech and animate the head; animated bodies to go with the faces and synthetic sound. A media object in its coded form consists of descriptive elements that allow handling the object in an audiovisual scene as well as of associated streaming data, if needed. It is important to note that in its coded form, each media object can be represented independent of its surroundings or background. The coded representation of media objects is as efficient as possible while taking into account the desired functionalities. Examples of such functionalities are error robustness, easy extraction and editing of an object, or having an object available in a scaleable form [Koenen02].

Severe distortions in a picture's contents may arise when there is only one P-frame between two I-frames. This is combated by placing so-called B-frames (B stands for bidirectional) between I-frames and P-frames. B-frames draw their information contents through referencing with the pictures which are transmitted both beforehand and afterwards. These pictures can be I-frames or P-frames. B-frames are also known as bidirectional referencing frames. A sequence of I, P and B-frames is called a GOP (Group of Pictures) [He02]. P-frames always have preceding I-frames or P-frames as reference, whereas B-frames use both preceding and follow-up I-frames or P-frames. The GOP size denotes the number of I, P and B-frames. GOPs may contain only one I-frame or one group of I (which is then similar to M-JPEG), P and B-frames. The MPEG standard permits GOPs of random length and composition, with a GOP always commencing with an I-frame (key picture). Maximum GOP lengths are laid down in order to guarantee good coding results by an encoder. The maximum length for NTSC is 18 frames and PAL 15 frames. The order within this limit is variable and can be directly influenced to improve the picture quality depending on the original (camera viewing angle, frequency of movement etc.). Wireless focussed video compression standards are becoming more common. Recent standards include MPEG-4 and H.26L [ITU-T02]. These standards are targeted towards 'reasonable' quality bit rates as low as 10kbps. The standards specify the structure of the bit stream, while leaving a great deal of flexibility in how the encoder creates a compliant bit stream. These wireless video coders are designed to perform under low spatial resolution, low frame rate, slowly moving and low-detail contents. Again, the main system design issues which arise in the wireless environment that these codecs must be aware of are data rate minimisation, complexity, power consumption and reliability.

2.3   Transmission Control Protocol/Internet Protocol

TCP/IP (Transmission Control Protocol/Internet Protocol) is a two-layered protocol. The *higher layer*, Transmission Control Protocol, manages the assembling of a message or file into smaller packets that are transmitted over the Internet and received by a TCP layer that reassembles the packets into the original message. The *lower layer* (network layer), Internet Protocol, handles the address part of each packet so that it gets to the right destination. TCP is an end-to-end protocol that is used for push and pull communication via a point-to-point TCP link. TCP is the primary transport protocol in the Internet (as well as Intranets and Extranets) suite of protocols providing reliable, connection-oriented, full-duplex streams and uses IP for delivery [Burks]. TCP's reference model differs from the OSI 7-layer model as can be seen in Figure 0-10.

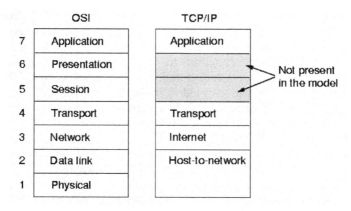

Figure 0-10: The differences between OSI and TCP/IP

The TCP/IP model traditionally has 4 layers (the Data Link Layer and the Physical Layer are combined into a single layer called the Host-to-Network Layer but it is also common to show these layers separately). There is no session layer or presentation layer in the TCP/IP model (if necessary, their tasks are made the responsibility of the application layer). The TCP/IP layers that exist are essentially the same as those in the OSI reference model. The *host-to-network* layer is responsible for sending bits across the network and for link error control and link flow control (i.e. data link layer and physical layer combined). The *Internet* layer switches packets around the network and places packets on (or removes them from) the network using a packet format called IP (Internet Protocol). The *transport* layer accepts data (and instructions) from the application layer and breaks the data up into packets. It also reassembles received packets into data streams and passes that data to the appropriate port number (e.g. port number 1066) and is also responsible for end-to-end flow control. Finally, the *application* layer contains all the higher-level protocols such as TELNET, FTP and SMTP that are used by applications.

2.3.1 Problems identified with TCP

TCP supports functionality such as adaptive retransmissions, deferred transmissions and delayed acknowledgements that can cause excessive overhead and latency for real-time applications [Schmidt97]. Likewise routing protocols like IPv4 lack functionality (such as packet admission policies and rate control) that can lead to excessive congestion and missed deadlines in networks and end systems [Rosenblum98]. A limitation for running multimedia applications over TCP/IP is the problem of delay. IP does not allocate a specific path or amount of bandwidth to a particular session. The resulting delay can vary widely and unpredictably, posing serious problems for real-time applications [Muller96]. For most organisations today, committing to IP is a given. The question is how far that commitment can be taken in an application sense. So far, the thrust of vendors in dealing with real-time IP has been to provide quality of service (QoS) control has been to implement forms of mapping IP flows to ATM virtual circuits to provide service guarantees or using an IP-based reservation protocol such as Resource ReSerVation Protocol (RSVP). Unfortunately, these solutions ignore the fact that TCP is not well suited for real-time applications.

TCP is a session protocol that sets up a virtual connection to the destination. Traffic flows in the session based on a window-size system that limits the amount of data a sender can send into the network. This structure is intended to reduce network congestion and buffer overflows as well as help speed slow receivers. TCP also provides error recovery, which is based on running a timer to detect a missing sequence-numbered data frame. It all adds up to reliable transport. In many cases, it also adds up to slowness. TCP session establishment is slow, particularly over long-delay networks such as the Internet. In addition, TCP error recovery relies on a time-out, which means that seconds may elapse before lost data is recovered. Finally, TCP has a "slow start" algorithm that limits window size to slow data flows when a new session is established or when a network error is detected. While this protects the network from congestion, it also interferes with continuous media application throughput. More real-time control can be provided using a special protocol in place of TCP. The Real-Time Protocol (RTP) [Schulz92, Liebl01, Johanson01], provides error recovery, flow control, and timing synchronisation between sender and receiver to improve real-time application performance. RTP and other Internet real-time protocols, such as the Internet Stream Protocol (ST-II) [Deering95, Ji03], focus not on the control-packet issue but on the efficiency of data transport. They are designed for communications sessions that are persistent and exchange a lot of data. RTP also includes the special RTP Control Protocol (RTCP), which can set up multipoint relationships. It is suitable for collaborative applications involving real-time video and audio. It has already been adopted by a number of voice-over-IP vendors to circumvent the limits of datagram transport of voice on the Internet and private IP networks.

2.4   Reliable Multicast Large-Scale Media Delivery

The Internet at present can only offer a best-effort network service due to varying link capacity and load. Therefore receivers in a large scale multicast group are going to experience different end-to-end delay variations and packet loss rates. The network scale and heterogeneity in available bandwidth complicate the design of network adaptive streaming multimedia multipoint applications [Tan01].   Forward Error Correct (FEC) can be used in the form of highly compressed low quality audio, which is piggybacked on normal audio packets. The decision on the level of FEC to use is made per source, based on receiver loss reports, and is tailored to cover for the average or highest requirements of the receiver group. This strategy is only good for a group observing similar loss rates. In a diverse group, receivers observing low losses are forced to receive useless redundant information, whereas receivers with very bad loss may not be covered [Wu02]. The variable network loss rates and perceived quality in different areas of a multicast distribution tree are a result of different link bandwidth availability and link load.

A single stream addressed to large whole groups cannot possibly cover the needs of all receivers. Instead the data rate and amount of redundancy has to be customised and separately distributed to problematic areas. It has been shown that sender driven schemes, which try to address the receiver heterogeneity problem, do not scale to large groups [Fulp01]. Any scheme that attempts to separately cover for the different needs of problematic receiver subgroups has to be receiver driven to scale [McCanne96]. Subgroups of co-located receivers in a multicast delivery tree, suffering from similar problems, should co-ordinate their efforts in improving reception quality. Furthermore their attempts should not affect reception for the remaining participants in the multicast session. Strict low delay requirements of real-time data distribution preclude solutions using retransmissions to achieve required reliability. The dynamic nature of the Internet delivery and membership model does not allow for manually configured static schemes that work around congested links. Proposals exist for integrating reliable multicast schemes into audio and video applications so that missing packets can be recovered from neighbours with better reception. This is achieved by trading off quality for delay, as any reliable multicast protocol has to request retransmission and wait for the repair. Although this may be acceptable in a real-time lecturing scenario, it becomes less useful with interactive communication. An additional undesirable side effect is that the operation of the reliable protocol creates extra control traffic.

Maxemchuk et al. [Maxemchuk01] propose a hierarchy of retransmission servers positioned around expensive or over utilised links. The servers operate a negative acknowledgement (NACK) based reliable protocol between them, and receivers use a similar scheme for requesting lost packets. Their proposal significantly improves reception quality but requires manual configuration of the retransmission servers. The STORM protocol [Xu97, Obraczka98] develops parent child relationships between participants of a multicast using an expanding ring search technique. Parents are chosen according to loss statistics, so that they have a good chance of receiving packets their children are likely to request. Streaming of stored data makes little sense unless browsing and selective playback is a requirement. For totally non real-time scenarios, a normal transport protocol and pre-fetch can be used to achieve perfect audio quality. TCP can be used in a single user scenario or a multicast congestion control protocol like RLC [Rizzo98, Gossain02] for multiple recipients.

Another approach is Self Organised Transcoding (SOT) by Kouvelas et al [Kouvelas98, Jamjoom01] which uses an adaptive mobile transcoder scheme to form groups out of co-located receivers with bad reception. A representative of the group is responsible for locating a suitably positioned receiver with better reception that is willing to provide a customised transcoded version of the session stream. The filtering site thus provides local repair to the congestion problem of the group, with minimal increase in stream delay. The data rate and redundancy level of the transcoded stream are continuously modified to adapt to the bottleneck link characteristics, using reception quality feedback from a member of the formed loss group. Network friendly congestion control of the real time multicast stream can thus be achieved.

2.5   Media Service Frameworks

Media service frameworks are middleware which are aimed at integrating multimedia services with service platforms. One such framework is PARLAY which is the service framework of the third generation partnership project (3GPP) [3GPP-03]. The use of the PARLAY APIs is proposed for the control of multimedia services [Parley03]. PARLAY is a forum established by key vendors and carriers with the goal to define object oriented APIs that allow third party application developers to access network resources in a generic and technology independent way. Pailer et al. [Pailer01] demonstrated that by coupling a service platform with communication services using the PARLAY API, superior control of the services is made possible, both from call control and mobility points of view. The service platform resides over a Session Initiation Protocol (SIP) [SIP, Handlwy99, Rosenberg02] server and thus a complete mapping of PARLAY events to SIP messages is required. Adoption of PARLAY is somewhat slow at this point since some carriers feel that they might lose the competitive edge to distinguish their services via this open API [Chen01].

The Internet is facilitating multiple forms of distributed applications, some of which employ application-level intermediaries such as Open Pluggable Edge Services [OPES]. The Open Pluggable Edge Services group's primary task is to define application-level protocols enabling such intermediaries to incorporate services that operate on messages transported by HTTP and RTP/RTSP. Open Pluggable Edge Services are services that would be deployed at application-level intermediaries in the network, for example, at a web proxy cache between the origin server and the client, that would transform or filter content. Examples of proposed OPES services include assembling personalised web pages, adding user-specific regional information to web pages, virus scanning, content adaptation for clients with limited bandwidth, language translation, among other applications. At the IP level, the participating intermediaries are endpoints that are addressed explicitly.

The emergence of ideas like middleware service providers supported by a group like OPES shows an industry trend to create value-added services on the network edge. The security model for such services involves defining the administrator roles and privileges for the application client, application server, intermediary, and auxiliary server. The data integrity model defines what operations are permitted by the content owner, and guarantees of content correctness can be made to the owner and viewers when content-related services are performed [Chen01, Rao01]

2.6 Reflection

Reflection allows us to separate the functional requirements of an application (what it does) from the non-functional ones (how it does it). It is based on the Meta-object Protocol (MOP) defined by Maes [Maes87]. The Java Reflection API (java.lang.reflect) puts particular emphasis on the use of Reflection in distributed systems by taking advantage of the network-centric capabilities of Java. The concept open implementation has recently been investigated by a number of researchers, most notably Kiczales et al at Xerox PARC [Kiczales96, Kiczales97, Hannemann02, Coady03]. The goal of this work is to overcome the limitations of the black box approach to software engineering and to open up key aspects of the implementation to the application. This must be achieved in such a way that there should be a principled division between the functionality they provide and the underlying implementation. The former can be thought of as the base interface of a module and the latter as a meta-interface [Rao91, Tanter03]. The role of reflection is then to provide a principled means of achieving open implementation. In a reflective system, the meta-level interface provides operations to manipulate a causally connected self-representation of the underlying implementation. According to Maes [Maes87], a system is said to be causally connected to its domain if the internal structures and the domain they represent are linked in such a way that if one of them changes, this leads to a corresponding effect on the other. Such a system has the benefits that, firstly, the self-representation always provides an accurate representation of the system, and that, secondly, a reflective system can bring modifications or extensions to itself by virtue of its own computation. Reflection allows a computational system to think about itself thus giving the possibility to enhance adaptability and to better control the applications that are based on it. For this, two different levels are defined: a base-level related to the functional aspects i.e. the code concerned with computations about the application domain, and a meta-level handling the non-functional aspects, i.e. the code supervising the execution of the functional code [Villazón00]. A reflective system therefore naturally supports inspection, and adaptation. Reflection enables applications to observe the occurrence of arbitrary events in the underlying implementation. Such an approach can be used to implement functions such as QoS monitors in a portable manner. Similarly, reflection allows applications to adapt the internal behaviour of the system either by changing the behaviour of an existing service (e.g. tuning the implementation of message passing to operate more optimally over a wireless link), or dynamically reconfiguring the system (e.g. inserting a filter object to reduce the bandwidth requirements of a communications stream). Such steps are often the result of changes detected during inspection.

A meta-object contains all the information of a base-level object and is able to control the execution and behaviour of the associated base-level object. The interactions between the base-level and the meta-level are governed by a Meta-object protocol (MOP) [Queloz99]. Thus, it is possible to manipulate base-level entities and even redefine how base-level entities are executed. Such architectures enable the development of highly flexible programs that can manipulate the state of their own execution. The combination of mobility and meta-level manipulations can provide a higher level of control and opens up interesting new possibilities. Reflection and meta architectures are means to achieve adaptability and mobile code technology gives extreme flexibility regarding distribution [Villazón00].

In networking, where applications can adapt the end-to-end path to particular requirements using code within intermediate proxies, reflection is an interesting mechanism that can be exploited to dynamically integrate non-functional code to an active network service. An increasing number of algorithms used in classical network models or classical distributed systems have been adapted to take into account benefits of reflective code such as Active Multicast [Lehman98] and Adaptive Routing [Willmott99]. Thus, the extreme flexibility of the active model is exploited, but on the other hand, the complexity of software design is increased. As a consequence, the composition of active services becomes very difficult and service designers integrate in the service code some aspects that are not directly related to the main functionality of the service itself. For example, tracing the activity of active packets and analysing how they interact with the different execution environments is a non-functional aspect that cross-cuts the original design of the service and that is often integrated in several parts of the software. Furthermore, the insertion of such code, in most cases implies stopping the service execution, integrating the modifications, recompiling and re-deploying the service over the network. Reflection gives a clean solution for structuring services in order to separate those orthogonal aspects. Dynamic integration of non-functional aspects is a notion that can be exploited in a context of adaptive media streaming. One of the major advantages of using reflection is that the information obtained from the execution of a service can be gathered and combined with similar information handed over several execution environments at the meta-level in order to enhance the overall service management, independently from the base-level service code [Villazón00].

2.7   Real Time Protocol and Application Level Framing

RTP [RFC1889, RFC 1890, Liebl01] is a real-time transport protocol supporting applications transmitting real-time data over unicast and multicast networks [Schulzr96a, Perkins03]. RTP is used in MBone audio/video tools in addition to numerous commercial implementations [McCanne95, Rosenberg99]. RTP services include payload type identification, sequence numbering, and time stamping. Delivery is monitored by the closely integrated control protocol RTCP. While RTP provides end-to-end delivery services, it does not provide all of the functionality typically provided by transport protocols therefore RTP generally resides on top of UDP to utilise its multiplexing services. End-to-end support includes multi-party functions, such as synchronisation of multiple streams, and reconstruction of streams based on timestamps. Sequence numbers allow the identification of packet positions in a stream that may   arrive out of order (e.g. to aid in video decoding). RTP bridges can act as synchronisation points along a path to transcode data into a more suitable format for the destinations that they serve [Miyazaki01].

Each RTP data packet consists of an RTP packet header and payload. The packet header includes a sequence number, a media-specific timestamp, and a synchronisation source (SSRC) identifier while the RTP control protocol (RTCP) provides mechanisms for data distribution monitoring, cross-media synchronisation, and sender identification. The control information transmission interval (sent to all participants) is randomised and adjusted according to the session size to maintain the RTCP bandwidth below some configurable limit. RTCP's primary function is to provide session feedback on data distribution quality, which is useful for diagnosing failures and monitoring performance, and can be adopted by applications for dynamic adaptation to congestion. Monitoring statistics such as Sender Reports (SR) include the sender's cumulative packet count and cumulative byte count while the Receiver Reports (RR) statistics include cumulative count of lost packets, jitter, short-term loss indicator, and round-trip time estimation time-stamps. RTP's design is based on the IP multicast group delivery protocol [Deering89] where data sources broadcast to a group's multicast address without knowledge of the actual group membership.

Real time applications such as adaptive audio or video conferencing tools are often based on RTP so that they can work in loaded networks so long as a minimal amount of bandwidth is available. Real time applications make use of RTP's support of intra and inter-stream synchronisation and encoding detection with RTP being frequently integrated into the application software rather than being implemented as a separate layer. In accordance with the ALF principle, the semantics of several RTP header fields are application dependent and several profile documents specify the use of RTP header fields for different applications (e.g. the marker bit of the RTP header defines the start of a talk spurt in an audio packet and the end of a video frame in a video packet) [Schulzr96b]. RTP will often be integrated into the application processing rather than being implemented as a separate layer oblivious to whether IPv6, Ethernet, ATM, or another communication channel is being used. It normally runs on top of UDP, using its framing services such as multiplexing and checksum.   The RTP packet header includes the identification of the synchronisation source (e.g. a combiner for audio data streams from participants in a conference) and the identifications of the sources contributing to the synchronization source.

The payload is a formatted Application Data Unit (ADU). An extension header may be added to allow individual implementations to experiment with new payload-format-independent functions without interfering with normal operations based on the regular header. The timestamp indicates the initial sampling time of the first byte in the RTP data packet. This sampling instant is derived from a clock with resolution adequate for the desired synchronization accuracy and for measuring packet arrival jitter. The initial value of the timestamp, as for the sequence number, should be random. Several consecutive RTP packets will have the same timestamp if they are generated at once, e.g., belong to the same video frame.

In 1990 Clark and Tennenhouse proposed a new protocol model called Application Level Framing (ALF), which explicitly includes an application's semantics in the design of that applications protocol [Clark90]. ALF promotes the breaking down of data into aggregates, which are meaningful to the application and independent of specific network technology. These data aggregates are known as Application Data Units (ADUs). The lower network layer should preserve the frame boundaries of ADUs so that applications are able to process each ADU separately and potentially out of order with respect to other ADUs. This ensures that ADU losses do not prevent the processing of other ADUs. In order to express data loss in terms meaningful to the application, RTP data units carry sequence numbers and timestamps, so receivers can determine the time and sequence relations between ADUs. As each ADU is a meaningful data entity to the receiving application, the application itself can decide about how to cope with a lost data unit (e.g. real-time digital video might choose to ignore lost frames, whereas FTP applications may request the resending of lost packets).

The RTP specification is designed to follow the principles of ALF and Integrated Layer Processing (ILP) [Abbott93, Mauve01] that advocates a tight integration of the network and codec processing into the application. Thus, an application can take advantage of the feedback information provided by RTP and adapt as needed to the condition and behaviour of the network. A corollary of this principle is that each application needs to have intimate knowledge of both network and codec behaviours. Existing RTP based audio/video applications such as vic (An MBone video conferencing tool) [McCanne95, Thorson01, Perkins03] use the principles of ALF and ILP to such an extent that RTP support is not easily extractable. There are no libraries so that RTP software cannot be reused for integration with existing applications, neither can new applications be designed to cleanly utilise the existing functionality thus the addition of new codecs and payload handlers is non trivial. ALF was later extended with a lightweight rendezvous mechanism based on IP multicast, aimed at receiver-based adaptation for real-time applications (e.g. audio and video conferencing). This is known as Light-Weight Sessions (LWS), and has been very successful in the design of wide-area, large-scale, conferencing applications.

2.8   Transcoding Proxies

A cellular wireless network consists of fixed based stations connecting mobile devices through a wired backbone network where each mobile device establishes contact through their local base stations. The available bandwidth on a wireless link is limited and channels are more prone to errors. It is argued that future evolution of network services will be driven by the ability of network elements to provide enhanced multimedia services to any client anywhere [Clark96, Harrysson02]. Future network elements must be capable of transparently accommodating and adjusting to client and content heterogeneity.

There are benefits to filtering IP packets in the wireless network so that minimal application data is carried to the mobile hosts to preserve radio resources and prevent the overloading of mobile hosts with unnecessary information and ultimately wasteful processing. Such filtering can be performed by a client-proxy-server architecture where mobile hosts execute the client application and the proxy is a machine in the wired portion of the wireless access network that intercepts and processes data packets travelling between client and server. This processing is transparent to the application servers in the wired network [Hunt98, Badrinath00, Joshi00, Matthur03].

A proxy is an intermediary component between a source and a sink, which transforms the data in some manner. In the case of mobile hosts, a proxy is often an application that executes in the wired network to support the host. This location is frequently the base station, the machine in the wired network that provides the radio interface. As the user moves, the proxy may also move to remain on the communication path from the mobile device to the fixed network. The proxy hides the mobile from the server, which thinks that it communicates with a standard client (i.e., a PC directly connected to the wired network) [Hokimoto96, Hokimoto97, Kammann02]. Typical examples of the use of proxies include:

- Transcoding high quality high-bandwidth colour streams into lower bandwidth monochrome for forwarding to the mobile device (e.g. Colour-space-scaling components reduce the number of entries in the colour space, for example from 24 to 12 bits, Gray-scale or Black-and-White).

- Image-scaling components resize video frames, which is useful to adapt a stream for devices with limited display (e.g. MPEG videos may be transmitted in any size while H.261 videos require predefined sizes such as CIF, QCIF or SQCIF).

- Retrieving large formatted files from a server (e.g. PDF, Ghostscript) and extracting a simple text version of the document for forwarding on to an impoverished mobile device (e.g. PDA) thus saving bandwidth and memory requirements on the mobile. Hence the proxy must convert an entire data store to a much simpler representation before relaying it to the client. Data insertion components can also be used to insert an image or a text in a video.

- Mixer components allow building a mixed video stream resulting from several input sources where the resulting video stream is an (NxM) matrix. Each element of this matrix results from an image-scaling adaptation of a particular stream. Multiplexors/Demultiplexors can also be used to aggregate/separate audio and video data in a multimedia stream (e.g. using an MPEG Multiplexor to merge an MP3 audio and MPEG-1 video in a MPEG2 stream. Duplicator components can be used to replicate an output media stream when a stream has different targets with different requirements.

- An application and its corresponding transcoder proxy reconfigure their structures when the machine configuration of a mobile computer is changed. E.g. if a memory card is removed, the size of an application buffer on the mobile computer may be reduced. This may require changing a buffer management policy (pre-fetch or on-demand) on a transcoder proxy.

Wireless links are characterised by relatively low bandwidth and high transmission error rates [Chakravorty02]. Furthermore, mobile devices often have computational constraints that preclude the use of standard Internet video formats on them thus by placing a mobile transcoding proxy at the Base Station (BS), the incoming video stream can be transcoded to a lower bandwidth stream, perhaps to a format more suitable to the nature of the device, and control the rate of output transmission over the wireless link [Rejaie00, Margaritidis00, Badrinath00, Joshi00].

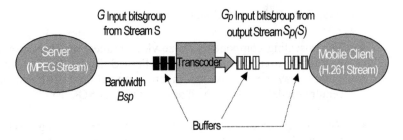

Figure 0-11 : A transcoding proxy

Figure 0-11 illustrates a scenario, where a transcoding gateway is configured to transcode MPEG streams to H.261. In the architecture, the transcoding gateway may also simply forward MPEG or H.261 packets to an alternate session (in both directions) without performing transcoding. Figure 0-12 illustrates locations in which intelligence about available network services may be placed. Clients may utilise this network knowledge to select the most appropriate server and mechanism in order to obtain appropriate content. As an alternative, this knowledge (and the associated burden) could be entirely or partially transferred to the individual servers or could reside inside the network.

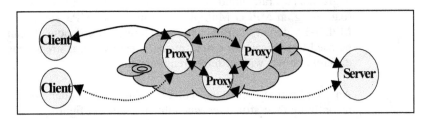

Figure 0-12 : Variations in Client-server connectivity

Image transcoding is where an image is converted from one format to another [Shanable00, Nagao, and Vetro01]. This may be performed by altering the Qscale (basically applying compression to reduce quality). This is sometimes known as simply resolution reduction. Another method is to scale down the dimensions of the image (spatial transcoding) [Chandra01] so reduce the overall byte size (e.g. scaling a 160Kb frame by 50% to 32KB). Another method known as temporal transcoding is where frames are simply dropped (this can sometimes be known as simply rate reduction). While another method may be simply to transcode the image to greyscale which may be useful for monochrome PDA's (again this transcoding process results in reduced byte size of the image or video frame). Recently there has been increased research into intelligent intermediaries [Tennen97, Tennen96, Scott97, Parr00, Rejaie00 and Bianchi00]. Support for streaming media in the form of media filters has also been proposed for programmable heterogeneous networking [Pasquale93, Yeadon96, Campbell97, Griwodz99, and Margaritidis00]. [Rejaie00, Elson01, Tran02] propose multiple proxy caches serving as intelligent intermediaries, improving content delivery performance by caching content. A key feature of these proxies is that they can be moved and re-configured to exploit geographic locality and content access patterns thus reducing network server load. Proxies may also perform content translation on static multimedia in addition to distillation functions in order to support content and client heterogeneity [Fox96, Gribble01, Yu01]. Repair heads servers [Xu97] in the network, can help with reliable multicast in a number of ways including performing ACK aggregation and retransmission. Another example is Fast Forward Networks Broadcast Overlay Architecture [fast00] where there are media bridges in the network, which can be used in combination with RealAudio [real02] or other multimedia streams to provide an application layer multicast overlay network. One could adopt the view at this time that "boxes" are being placed in the network to aid applications.

2.9   Mobile Communications

Mobile phone technologies have evolved in several major phases denoted by "Generations" or "G" for short. Three generations of mobile phones have evolved so far, each successive generation more reliable and flexible than the previous. The first of these is referred to as the first generation or 1G. This generation was developed during the 1980s and early 1990s and only provided an Analog voice service with no data services available [Bates02].

The second generation or 2G of mobile technologies used circuit-based digital networks. Since 2G networks are digital they are capable of carrying data transmissions, with an average speed of around 9.6K bps (bits per second). The three standards involved in 2G networks are Time Division Multiple Access (TDMA), Code Division Multiple Access (CDMA) in America and Global System for Mobile (GSM) used in both America and Europe [Garg99, Stuckmann02]. Because 2G networks can support the transfer of data, they are able to support Java enabled phones. Some manufacturers are providing Java 2 Micro Edition (J2ME) [Feng01] phones for 2G networks though the majority are designing their Java enabled phones for the 2.5G and 3G networks, where the increased bandwidth and data transmission speed will make these applications more usable [Hoffman02].

"2.5" is an acronym which represents various technology upgrades to the existing 2G mobile networks. Upgrades to increase the number of consumers the network can service while boosting data rates to around 56K bps. 2.5G upgrade technologies are designed to be overlaid on top of 2G networks with minimal additional infrastructure where examples include General Packet Radio Service (GPRS) [Chakravorty02] and Enhanced Data rates for Global Evolution (EDGE) [Furuskar99]. These are packet based and allow for "always on" connectivity. The third generation of mobile communications (3G)[1] is digital mobile multimedia offering broadband mobile communications with voice, video, graphics, audio and other forms of information. 3G builds upon the knowledge and experience derived from the preceding generations of mobile communication, namely 2G and 2.5G although 3G networks use different transmission frequencies from these previous generations and therefore require a different infrastructure [Bates02]. These networks will improve data transmission speed up to 144K bps in a high speed moving environment, 384K bps in a low-speed moving environment, and 2Mbps in a stationary environment. 3G services see the logical convergence of two of the biggest technology trends of recent times, the Internet and mobile telephony [Barbounakis00]. Some of the services that will be enabled by the broadband bandwidth of the 3G networks include:

- Downloadable and streaming Audio and Video.
- Voice Over Internet Protocol (VoIP).
- Send and receive high quality colour images.
- Electronic Agents – are self-contained programs that roam communications networks delivering /receiving messages or looking for information or services.
- Downloadable Software – Potential to be more convenient than conventional methods of distributing software as the product arrives in minutes.
- Capability to determine geographic position of a mobile device using the Global Positioning System (GPS) [Barnes03].

3G will also facilitate many other new services that have not previously been available over mobile networks due to the limitations in data transmission speeds. These new wireless applications will provide solutions to companies with distributed workforces, where employees need access to a wide range of information and services via their corporate intranets, when they are working offsite with no access to a desktop [Smith01].

---

[1] http://www.3gnewsroom.com

NTT DoCoMo launched "i-Mode" in 1999, a multimedia data service with Internet access [Vacca99]. By the beginning of March 2001, NTT DoCoMo announced that they had exceeded their target of 20 million i-Mode sub-scribers. This success was mainly driven by the wide variety of applications which where available for the NTT DoCoMo customers. This packet-based service was specifically designed for mobile phones and was the world's first Web browsing interface over a wireless device. i-Mode is an overlay over NTT DoCoMo's ordinary mobile voice system. It is an end-to-end solution consisting of a transmission system; mark-up language and smart phones built upon a personal digital cellular-packet (PDC-P) technology with an air interface and a time division multiple access (TDMA) system with a 3/6 timeslot-per-frame structure, depending on the traffic load. With the radio interface, the packet data over PDC's TDMA channels are implemented in a method, which is similar to the general packet radio service, which is used in the global system for mobile communication and the packets are inserted into fixed-length timeslots.

2.9.1   Mobile IP

Mobile IP [Perkins98b] is an extension to IP, which allows the transparent routing of IP datagrams to mobile nodes. In Mobile IP each host has a Home Agent (i.e. a host connected to the sub network the host is attached to). The home agent holds responsibility for tracking the current point of attachment of the mobile device so when the device changes the network it is connected to it has to register a new care-of address with the home agent. The care-of address can be the address of a Foreign Agent (e.g. a wireless base station node) that has agreed to provide services for the mobile or the new IP address of the mobile (if one is dynamically assigned by its new network). Traffic to mobiles is always delivered via home agents, and then tunnelled to the care-of address. In the case of a foreign agent care-of address, traffic is forwarded to the mobile via the foreign agent. Traffic from the mobile does not need to travel via the home agent but can be sent directly to (previously) correspondents. Correspondent hosts do not need to be Mobile IP enabled or even have to know the location of the mobile if a home agent acts as an intermediary, thus forwarding of packets to the current address of the mobile is transparent for other hosts [Dixit02]. The home agent redirects packets from the home network to the care-of address by creating a new IP header, which contains the mobile's care-of address as the destination IP address. This new header encapsulates the original packet, causing the mobile node's home address to have no effect on the encapsulated packet's routing until it reaches the care-of address.

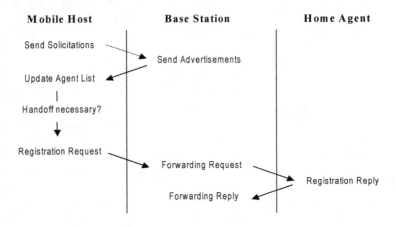

Figure 0-13 : Mobile IP Handoff Algorithm

When the mobile leaves the service area of its current foreign agent and registers with a new foreign agent, the home agent must be informed about the change of address. In the process of handoff, the mobile may lose connectivity for a short period of time. The handoff commences with a Registration Request from the mobile to the base station, which forwards the request to the home agent of the mobile which then updates the care-of-address (COA) of the mobile and installs an encapsulator to tunnel IP packets to the mobile via the base station. The home agent then sends a Registration Reply message to the mobile via the base station stating that handoff was successful. The base station assumes responsibility as the foreign agent for the mobile. The handoff algorithm itself is kept very simple as illustrated in Figure 0-13. Mobiles send registration requests upon reception of a beacon message from a base station, and then start to use that base station as a foreign agent. This results in a dropout until the new connection is established although the mobile could still communicate with the rest of the network over its current foreign agent. The Mobile IP architecture is more suited to mobile's that change their point of attachment as opposed to fast moving hosts as registering a care-of address with the home agent can lead to over-head and high delays, which in turn cause decreased protocol performance. While not generally available in wide-area Internet paths, Mobile IP can readily be deployed in enterprise wide networks today. All major operating systems support IP Multicast and multicast routing support in network devices is mature and commercially available thus Mobile IP can easily be deployed over this multicast structure.

2.10   Commercial Streaming Systems

Media streaming is transmission through the network at the play-out rate. Bandwidth limitations, variable delays, and congestion in the packet network require rate-adaptive mechanisms, and buffering en route or at the receiver to smooth arrival perturbations.   These mechanisms operate between a media server providing the stream, and one or more media clients running on the receiving device. Streaming implies that both live media and stored files are downloaded at the playing rate. The major commercial streaming media leaders at present are RealMedia from Real Networks Inc., Microsoft's MediaSystem platform and QuickTime from Apple Corporation. These systems are discussed here.

2.10.1   The RealMedia Streaming System

RealSystem developed by RealNetworks[1], Inc. is a widely-used commercial media streaming system.   RealNetworks, a pioneer and leader in the media streaming industry, was an early adaptor of RTP and co-authored RTSP and SMIL. The RealSystem is a complete streaming infrastructure including authoring tools, digital encoding software, the RealServer, and the client system realised in software-based media players such as RealPlayer G2 and later versions. The interactive process for unicast RealMedia streaming in the Internet is illustrated in Figure 0-14.

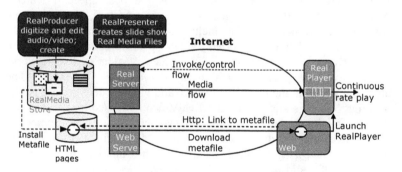

Figure 0-14: Basic functions of the RealMedia system

---

[1] http:www.real.com

Analog audio and video streams are fed to the RealProducer application, which digitises them and facilitates labeling and classification of time segments. They are stored, conventionally on the editing computer's hard disk, and are uploaded to the store associated with the RealMedia server where they are cached for current use. Different versions of the RealServer are designed to serve different numbers of simultaneous users, ranging from about 100 to many thousands. A stored segment is represented by a text metafile on a Web server, available through an HTML page. A user links to the metafile from an HTTP client. When the metafile is requested by the client, it is downloaded. There it executes and activates the RealPlayer client, causing it to request a download streaming of the relevant media file from the RealServer. This is an RTSP interaction. As the data stream begins to come in, the RealPlayer buffers it until enough contents are in the buffer to initiate play-out.

RealSystem can accommodate standard audio and video formats. It licenses several third party proprietary formats, and has its proprietary formats that it claims outperform competitors such as RealMedia (.rm), RealAudio (.ra), RealPix (.rp), RealText (.rt) etc. A RealMedia (.rm) file is an audio or video clip encoded by one of the applications RealVideo Encoder (version 8.0 at time of writing), RealEncoder or RealPublisher. Files created with RealEncoder or RealPublisher can contain multiple streams, including audio, video, image maps, and events. Video files can contain audio and video or video only and can be played by a RealPlayer client, as can audio only files. The audio codecs supported include RealAudio 8 (<96kbps, monaural and stereo), ACELP coder (5kbps, 6.5kbps, 8.5kbps, 16kbps, speech-oriented), ATRAC3 coder (>96kbps, high quality music-oriented), MP3 (for Real Jukebox) and others plug-ins including Liquid Audio, A2B, MIDI, WMA etc. Some supported video codecs are RealVideo 8 (RealNetworks proprietary), G2 Video (older RealNetworks proprietary) and plug-ins such as MPEG-4, On2 and more). The algorithms used in the proprietary audio and video codecs are not public information. It is known that the RealVideo coder exploits transform-domain spatial coding, motion-compensation coding and subjective error concealment as does MPEG, and uses MPEG-like I (Intra)-frames for seek/fast forward/rewind needs. The interleaving used for protection of audio packets could not be used in video because of the larger packet sizes, and was replaced by forward error correction and subjective error concealment mechanisms.

2.10.2   Microsoft Windows Media

Microsoft Windows Media 9 Series[1] (released in early 2003) is a comprehensive solution to media encoding, serving, delivery, and play-out of media. The platform includes Windows Media Encoder Series, Windows Media Services 9 Series, Windows Media Player 9 Series, and the Windows Media Audio and Video 9 codecs. Some of the attributes claimed for this platform, including proprietary audio and video encoding formats, are:

- Fast Start and Fast Cache features for rapid startup and *always-on* play. The server begins with a higher-rate burst, assuming the bandwidth is available, to quickly bring the buffer occupancy to an acceptable starting level. It also supports streams at different bit rates, as in the RealSystem. An added capability is support of multiple languages and variable speed playback without changing the pitch and tone of the original audio. It also provides a smart jukebox capability for dynamic music play lists and synchronised display of lyrics.

- Claimed compression improvements of 20% for Windows Media Audio 9 and 15-50% for Video 9 compared with the previous generation Windows Media Audio and Video 8. Speech and music content can be mixed and delivered at sub-20Kbps rates. Streaming is possible at rates needed for both dialup and broadband connections. The professional Audio 9 encoder also supports multi-channel 5.1 surround sound.

- Digital video including high definition, supporting video frame sizes of 1280x720 with software decoding and 1920x1080 with hardware graphics chips. The professional Video 9 encoder claims three times the compression efficiency of MPEG-2. A server of increased capacity that accommodates advertising insertion and digital intellectual property rights management and integration of the player into a wide range of wired and wireless consumer appliances, including car stereos and DVD players.

Streaming is done with both proprietary protocols and RTSP. RTSP is used for delivering content as a unicast stream, but only when delivering the stream to Windows Media Player 9 Series.

---

[1] http://www.microsoft.com/windows/windowsmedia/default.aspx

2.10.3   Quicktime

QuickTime[1] is an integrated collection of technologies similar to Microsoft and Real Networks media platform offerings. Quicktime supports Formats including AVI, BMP, GIF, JPEG, MPEG, WAV, PhotoShop, FlashPix and DV. Video Compressors include Cinepak, Intel Indeo Video 3.2 and 4.4, Photo JPEG and Sorenson Video 1 and 2 and sound compressors include IMA 4.1, QDesign Music2 and QualComm Pure Voice. There are predefined settings for specialised compression formats ranging from streaming 20kbps music to 2x CD-ROM using Sorenson video. Quicktime includes HTTP and RTP/RTSP streaming capabilities with versions of the popular Sorenson and QDesign codecs. Quicktime also supports export formats including AVI, and a brushed aluminum appearance with audio EQ display and controls. Other features include automatic selection of the best-quality movie for a given Internet connection, processor speed or spoken language; Internet decompressors such as H.263, GSM, MS DVI, RTP, DVI, MPEG-1 Layer 3; a dynamic Internet installer where a component installation updates a system automatically; media import and export for PNG, TIFF, TARGA, MacPaint, Macromedia Flash, and FlashPix formats.

Fast start streaming which is built into QuickTime downloads the file from the server to a hard drive and begins playing it once enough content has been received to provide uninterrupted viewing. QuickTime fast start enables users to begin watching videos before the download is complete. QuickTime also provides the option to save and replay the file as many times as the client wishes. The RTSP protocol does not download the movie to the hard drive. Instead, the computer receives a steady flow of information displayed in the web browser and is discarded once the customer has viewed it. To help offset minor network interruptions, a cache is filled and maintained with three to 10 seconds of data. In February 1998, the International Standards Organization adopted the use of the QuickTime media format as the basis of the MPEG-4 specification because of its flexible format and ability to stream media across different network protocols.

---

[1] http://www.apple.com/quicktime/

2.11   Chapter Summary

The research problem addressed in this book covers a diverse range of technologies. This chapter has sought to provide an introduction to the various technologies investigated as part of the solution domain. Networking has been examined at a low-level micro-protocol level and again from a high-level middleware framework viewpoint. Meta-level manipulations through the principle of Reflection were introduced as a promising way to achieving dynamic configuration, especially because of the resulting openness, which enables the programmer to customize the structure of the system. Software engineering principles have been touched on with regards correct principles of protocol stack layering and again with the application level framework paradigm. Issues regarding mobility were discussed among others. The following chapter provides a research background of many of these key areas involved and paves the way towards the new framework.

3   Need for a New Paradigm in Middleware

A solution to problems discussed previously, which is addressed by this work, is to provide a quality aware media transcoding middleware which allows clients to select among a range of media formats and within each media stream to convert these multimedia streams to more appropriate formats on the fly if necessary. The converting process is known as transcoding, which means converting multimedia streams from one format to another format. To recap - the explosive growth of the Internet and mobile computing has brought to light two main problem areas in delivering high quality multimedia streams to moving targets. The first problem area is heterogeneity of client devices and their network connections with client devices varying from desktop PCs, notebook computers, PDAs to mobile phones, with their capabilities also varying along many axes, including screen size, colour depth and processing power [Fox96, Hughes01, Upadhyaya02]. Furthermore, they may connect to the Internet via different networks, such as wired LAN, wireless LAN or wireless WAN. The second problem area is mobility of clients, which can be moving while they are accessing multimedia streams thus allowing network connections to change from time to time, ranging from a good (i.e. high throughput error free) network to a congested network [Upadhyaya02, Hansmann03]. The two problem areas described above make it difficult for a multimedia server to provide a streaming service, which is appropriate for every client in every situation. The focus of this body of work is the problem of delivering optimal multimedia streams to a 'heterogeneous sea' of memory & resource constrained mobile devices experiencing fluctuating network conditions. What follows are some of the key issues, which have been identified as barriers to the above and which form the core group of problems addressed by the middleware which contribute to an overall framework for improving delivery of media to mobile devices.

Computers communicate through the use of a common set of protocols that define the set of rules to be adopted for the duration of the communication. Middleware protocols stacks have traditionally been monolithic chunks of code where all data regardless of whether it is a continuous stream of bits with strict time dependencies between those bits, or the packets comprising an asynchronous message are sent through the same stack. The nature of the data is not considered and therefore there is no room for optimisation of the code to create a more efficient service [Schaphorst96]. Protocol functionality can exist which is never invoked leading to unnecessary stack functionality, which ultimately can lead to inefficiency in stack execution [Talley97, Braden02]. These redundant stack components also lead to extra amounts of memory being utilised which in the case of memory constrained mobile devices result in the sacrifice of precious processing cycles and ultimately increased power consumption [Kravets01].

Information travelling over wireless networks is prone to increased error as opposed to data over a wired local area network thus an argument exists for protocols tailored to the nature of each underlying network medium [Yamamoto01, Hansmann03]. A protocol such as TCP can be used to transport data over this medium; but TCP applies a rate controlling mechanism, which halves the current throughput upon detecting congestion. Congestion is detected by recording lost packets but losses are likely to occur on wireless networks due to error rather than network congestion thus the TCP congestion mechanism is inappropriate [Kojo97, Dixit02]. A generic monolithic protocol stack, which contained mechanisms to cope with every use case could be developed but this would lead to a solution with a large degree of redundancy as many functions would not typically be called. Additionally the amount of user space memory required to implement this solution would result in the exclusion of memory-constrained devices.

Internet-connected mobile devices such as mobile phones, PDAs, and mobile information appliances are increasingly becoming important tools in our daily lives. The ability to access information in real-time and in real-life circumstances has enabled us to work more efficiently. To fully harness mobile commerce's potential, there is a need to integrate wireless devices with the rest of the Internet [Yuan02]. In particular, wireless devices should be able to access the vast amounts of multimedia on backend but current and next generation wireless devices, have limited processing power in comparison to their desktop counterparts [Lee00, Hughes01]. This seems likely to be the trend for the near future. Multimedia has strict real-time requirements with regards reception; processing and display of media therefore large reservoirs of multimedia are likely to be inaccessible to these memory constrained devices for the foreseeable future [Han01, Upadhyaya02].

Another problem exists in the heterogeneity of the mobile devices, which have different capabilities ranging along many axes including screen size, colour depth and processing power [Fox96, Garg99, Fong03]. Processing power may range from powerful full specification laptop PC's, to low powered PDA's while other devices will be capable of displaying full colour 1024x800 and others only managing black and white 100x60 screen resolution. Differences also exist in each devices network connection bandwidth. Some devices will be connected to T1 or broadband channels; some may connect using 802.11b, ISDN or 56K modems while other clients may connect using first generation GSM networks. Middleware to date has tended to provide no support for these issues but other approaches to the problem have involved sending the lowest common denominator stream to all receivers, but this penalises the more powerful mobile clients that receive far below their true capabilities.

Quality of Service is an end-to-end concept where every box, router and bridge must support QoS thus Quality of Service at present is not for use over the Internet as it requires end-to-end support [Trimintzios01]. TCP/IP networks are essentially best-effort conduits in which packets are sent wholesale, without waiting in line for each other. Packets routinely collide and hit congestion, causing tiny delays and often requiring resends. Those delays, usually measured in milliseconds, are meaningless with data traffic but wreak havoc with time sensitive multimedia traffic. There is no clear picture of what it will take for better QoS to be deployed in enterprise networks through middleware, and demand is not forcing fast solutions, but solutions to many QoS problems can be solved using application level QoS mechanisms and to date much of the enterprise middleware available ignores or implements poorly, priority assignments to media. There exists a need for autonomous and time-critical behaviour in next-generation applications necessitating more flexible system infrastructure components that can adapt robustly to dynamic end-to-end changes in application requirements and environmental conditions. Next-generation applications will require the simultaneous satisfaction of multiple QoS properties, such as predictable latency/jitter/throughput, scalability, dependability, and security. Applications will also need different levels of QoS under different configurations, environmental conditions, and costs, and multiple QoS properties must be coordinated with and/or traded off against each other to achieve the intended application results [Schantz01]. Improvements in current middleware QoS and better control over underlying hardware and software components as well as additional middleware services to coordinate these will all be needed.

There is currently a notable lack of adequate development frameworks to cover the above requirements necessary for dynamic mobile communications, as classical methods for distributed systems are not sufficient due to their static character. Implementing protocols from scratch is a complex and time-consuming task. Frameworks are designed to ease the task of developing new protocols. They achieve this by providing a library of basic protocol functions and templates for implementation of new protocols (e.g. the X-Kernal [Peterson91], STREAMS [Unix90], Conduits [Huni95], MobiWeb [Margaritidis00] and Globus [Foster01]). The weakness with these protocol frameworks is that they are not portable while other frameworks have for the most part been proprietary in nature and lacking in any standard API to enable new mechanisms to be deployed at a future time to cope with additional use cases. Many mobile manufacturers at present including Motorola, Ericcson and Nokia have devices in the marketplace, which support the Java 2 Micro-Edition (J2ME) framework[1]. The trend towards the porting of implementations of Java on mobile devices is expected to continue[2] as it is an ideal language for protocol implementation due to its extreme portability and support for modular programming in an object oriented fashion. A study by SRI Consulting Business Intelligence in April 2003, titled 'Programmable Mobile Phones: The Battle over Platform Software,' [3]suggests that many manufacturers and developers are shifting their platform choices away from leading mobile platforms like Microsoft, Palm, Qualcomm and Symbian, and are instead favouring Java. According to the study's author, Michael Gold, a senior engineer in the Digital Futures Program at SRIC-BI, this recent trend is due in part to licensing terms, which for Java are more favourable and cheaper than Microsoft, Palm, Symbian, and others. According to Gold, Java phones don't require as much memory and processor power, which enables longer battery life. And because Java licensing terms are less expensive, manufacturers can more cheaply market a Java device. An estimated 50 million Java handsets are currently on the market and shipments of Java handsets exceeded PDA shipments in 2002, said Gold, adding that one out of every ten cell phones in the world will be Java phones by the end of the year. Hence, what is required is a portable framework that supports dynamic QoS adaptation in a changing environment, ideally containing small footprint so as to enable the framework to be run on the next generation of portable devices.

3.1   A New Middleware Framework

A paradigm is proposed whereby media are transported through an optimised stack constructed solely for that medium. Applications ship media through micro stack objects and the protocol graph is reconfigurable in cases where the operational environment changes. Configurable end-system light-weight protocols support a wide range of application requirements and increase protocol performance and systematic rapid application development by decreasing protocol complexity. The middleware supports QoS inspection at various points, and enables applications to dynamically adapt the behaviour of the underlying platform in response to QoS changes.   A key requirement

---

[1] http://e-www.motorola.com/webapp/sps/site/homepage.jsp?nodeId=03M0ym4sDZx
[2] http://www.javamobiles.com/
[3] http://www.internetnews.com/dev-news/article.php/1598081

here is that the mobile adaptation methods ensure the delivery of the expected communication quality without wasting valuable radio spectrum resources and with practical distribution of the processing load.

Proxies located at the home agent of each mobile client can also offload processing for memory-constrained devices while also assuming responsibility for activities such as message queuing and forwarding, access control to streaming sessions, message encryption and protocol translation among others. Proxies relieve clients of most of the work of maintaining state information about active sessions, filters and general system monitoring facilities and protocol stacks have been adopted to simplify the interoperability process, by allowing code for dynamic stacks to be written once and placed on repositories where they can be shared allowing thin clients to download new protocol stacks. A crucial benefit is that the actual code footprint for all the stack elements can be minimised with only the required stacks being actually in place on the client machine (rather than the entire potential library) thus allowing protocol stacks to fit on limited memory devices that might otherwise not have been possible.

Chameleon offers a hybrid approach to the problem of the burstiness of VBR traffic where bandwidth is dynamically allocated to match the varying traffic load so that prioritised media streams are able to 'borrow' bandwidth from lower priority streams. This will lead to optimal performance over a range of offered loads both in micro and macro-cellular environments. Heterogeneous clients are provided for by supplying various qualities of data and sending these media to separate multicast groups so that clients can select a Quality of Service (QoS) in accordance with available resources. This mechanism offers movement between multiple multicast media groups and transcoding of media within groups (no need to join a separate group). This book presents two innovative QoS mechanisms which are named PQT and SQT where a Primary Quality Transformation (PQT) technique, can assume responsibility for coarse grain adaptation decisions by moving between multicast groups upon violation of group bandwidth limits. Secondary Quality Transformation (SQT) complements PQT by assuming responsibility for responding to quality fluctuations within each group. Both techniques can be optimised to work with priorities being assigned to differing streams within a link in order to allow different streams to be rate controlled according to application-implied importance. Mobile transcoding proxies also tackle the problem of heterogeneous group communication by being dynamically loaded and activated within the network to provide for individual services within a heterogeneous multicast group. To support the PQT technique, a simple form of traffic differentiation is introduced into the network by designating streams which are queued (during congestion) as high, normal, medium, or low. Using Priority Queuing, all high-priority traffic is serviced first, then normal, and so on (e.g. Audio, then Video etc).

One key idea behind the middleware is the uniform abstraction of services as well as device capabilities via proxies as the application-programming interface. Consequently, the middleware delivers requests to either device services in the middleware or transport protocols. The provision of alternative communication models with respect to the transactional pattern results in a middleware that provides the synchronisation independent of the underlying protocols. Our approach is inspired by micro-kernels as they were introduced into the realm of operating systems [Rashid89, Tanenbaum91] and had some first applications in the middleware area as well [Puder00, Roman01b]. The middleware accepts invocations which are typically composed of a source and a target address, an operation with parameters, and additional information concerning the handling of the invocation. The middleware dispatches the invocation to a local service, a local device capability or a transport plug-in, which transports the invocation to a remote middleware. Transports that receive an invocation submit them to the middleware to initiate the dispatching to the corresponding local service or device capability. Invocations can be either generated by proxies, representing a service or a device capability, or manually by the application programmer, e.g. like the request object in the dynamic invocation interface in CORBA. The requirement for uniform access of device capabilities as well as remote services can be easily established by this approach. The middleware allows the flexible integration of new transport plug-ins and device capabilities by simply registering a new entity, which accepts an invocation. This allows the provision of access to all features available on resource-rich computer systems.

A key characteristic of the framework is that it should evolve and adapt to changing environments similar to living things in their struggle for survival. This can be achieved through a layered Java framework providing an API for a set of middleware services that it abstracts building upon multimedia, mobility, middleware and reflective frameworks to provide a union of these service API's in a streamlined API. The middleware is implemented in Java allowing it to be deployed on all platforms for which a Java VM exists including specialised Java processors [Jstamp02]. The proliferation of end-systems besides classical computers capable of executing Java, such as cell-phones or PDAs, and the aforementioned embedded systems make Java a suitable starting point providing a uniform abstraction for our middleware. The portability of Java make it an ideal middleware language for future systems overcoming the non-portable limitations of the previously discussed protocol frameworks. A simplified programming model (reduced instruction set) with a high level control API allowing elements to be optimised in software. The benefit of our micro-protocol stack approach compared to existing middleware platforms is the minimal footprint needed for a basic configuration which qualifies it for embedded systems as well as the extensibility providing the means to use features of more sophisticated computers. The configurability that reflective middleware typically provides is also supported by us. A major difference to existing middleware platforms is the support of different communication models, such as RPC or events with different synchronization semantics, by the middleware, which allows these communication models over a variety of different interoperability protocols.

The purpose of this new middleware infrastructure is to deliver optimal multimedia streams in heterogeneous environments. There exists a lack of adequate development frameworks to cover the characteristics of dynamic mobile communications, as classical methods for distributed systems are not sufficient due to their static character. The weakness with many existing protocol frameworks is that they are not portable while other frameworks are for the most part proprietary in nature and lacking in any standard API to enable new mechanisms to be deployed at a future time to cope with additional use cases. Java is an ideal language for protocol implementation due to its extreme portability and support for modular programming in an object oriented fashion. Hence, what is required is a portable framework that supports dynamic adaptation in a changing environment, which is systematic, yet, axiomated on practical experience and contains a small footprint so as to enable the framework to be run on the next generation of portable devices. Thus, the Java middleware presented here allows a static or mobile client on a network to request a multimedia stream from a selection of pre-encoding media residing at separate multicast group addresses. Priorities can be assigned to specific media such as audio and transcoders may further transcode each media stream to a more appropriate format for the client should the need arise. The middleware will transparently invoke the appropriate transcoder and build a support chain from the server to each client. Additionally, protocol stacks may be reconfigured at run-time using reflection in order to provide 'best-fit' configurations for optimal transport of media.

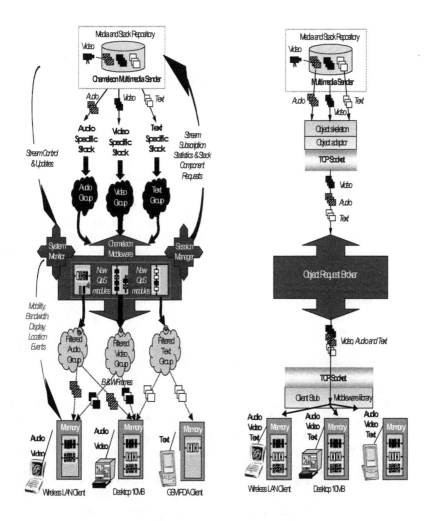

Figure 0-15: Chameleon Middleware

Figure 0-16: ORB Middleware

Figure 0-15 provides a comparison between the proposed middleware architecture with a generic ORB based middleware (Figure 0-16) such as one modelled on the CORBA architecture. The CORBA architecture is illustrated here as an example of a monolithic protocol stack composed middleware as opposed to a dynamically configurable lightweight protocol stack middleware like Chameleon. The following section recaps on problems hindering streaming of media to mobile clients and provides an outline of tests, which validate proposed solutions in each of the problem areas.

4   The Chameleon Framework

An object-oriented framework is a skeleton implementation of an application in a particular problem domain composed of concrete and abstract classes, which provide a model of interaction among the instances of classes defined by the framework. An important characteristic of a framework is that the methods defined by the user to tailor the framework will often be called from within the framework itself, rather than the user's application code thus the framework often plays the role of the main program in co-ordinating and sequencing application activity. This inversion of control gives the framework the power to serve as extensible skeleton with the methods supplied by the user tailoring the generic algorithms defined in the framework for a particular application [Steyaert98, Schmidt03]. The data flow [Steinmetz92, Steinmetz96, Eide01] principle has been adhered to for implementing Chameleon where the class hierarchy derived from the data flow principle consists of source classes, sink classes, intermediate processing classes, and connecting objects. Applications are built by forming a data flow path of connected objects, which are instantiated from appropriate classes.

Chameleon is a Middleware Framework [Schantz01] (so named, as it adapts 'automatically' in an 'optimal' manner to a fluctuating network environment delivering streaming media), which supports reconfigurable dissemination oriented communication [Parr-Curran98a, Parr-Curran98b, Parr-Curran98c, Parr-Curran99, Parr-Curran00]. Chameleon fragments various media elements of a multimedia application, prioritises them and broadcasts them over separate channels to be subscribed to at the receiver's own choice. The full range of media is not forced on any subscriber, rather a source transmits over a particular channel, and receivers, which have previously subscribed (to the channel), receive media streams (e.g. audio, text and video) with no interactions with the source. Clients are free to 'move' between differing quality multicast groups in order to receive the highest quality (or move to a lower quality group for the greater good of minimising network congestion. This is known as Primary Quality Transformation (PQT). In addition proxies offload intensive computing on behalf of clients and dynamic reconfigurable abilities allow new components to be slotted into live systems. The new components can perform additional transcoding on streams within each group. This is known as Secondary Quality Transformation (SQT). PQT and SQT provide a rich set of features for the optimal reception of multimedia flows. Chameleon is packaged with a core API and a set of Java template classes. The object-oriented design process produces a hierarchy of classes, from which a collection of objects is instantiated to build a particular application. A partial illustration of key chameleon components is illustrated in Figure 0-17.

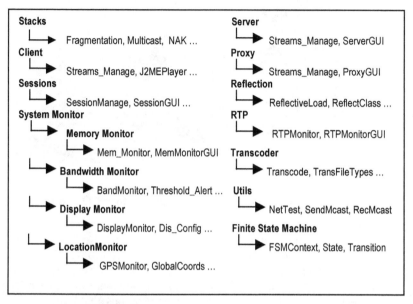

Figure 0-17 : Chameleon Classes

Chameleon (Figure 0-18) addresses the network congestion and heterogeneity problem by taking into account the differing nature and requirements of multimedia elements such as text, audio and video thereby creating tailored protocol stacks which distribute the information to different multicast groups allowing the receivers to decide which multicast group(s) to subscribe to according to available memory, display resolutions and network bandwidth availability.

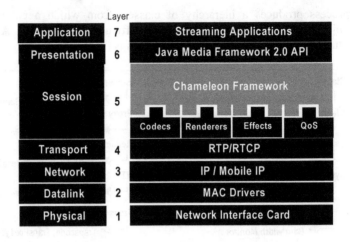

Figure 0-18 : Chameleon in comparison to the OSI model

Chameleon supports dissemination of multimedia from a source to multiple destinations but end-to-end closed-loop control can be difficult and cumbersome with multiple receivers, as the slowest receiver will impede the progress of the others. It can be argued that tight, closed-loop, end-to-end control is inappropriate for applications that expect a large number of receivers having different capabilities interconnected through networks providing different QoS. Instead, the decision taken here has been to adopt an alternative approach that relies on very loose coupling between the source and the receivers, i.e. an open-loop approach, more suited to real-time continuous media. A more detailed architectural view is provided in Figure 0-19.

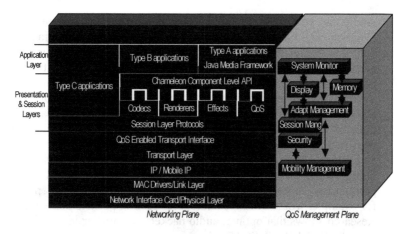

Figure 0-19 : Chameleon Detailed Layered Architecture View

In Chameleon, applications interface with a QoS (where QoS is defined as any non-functional property of an application) and mobility aware protocol stack through a set of interfaces each addressing one of the above application types. Type *A* applications specifically written with the Java media framework in mind which allows the highest level of integration with the middleware. Type *A* applications can exploit all aspects of the middleware including media specific stacks, system adaptation, session management, mobility etc. Type *B* applications are not written for the Java media framework but can still access many parts of the middleware to exploit aspects such as adaptation and mobility. Type *C* applications do not possess the ability to adopt media specific protocols stacks but some aspects of dynamic reconfiguration of stacks is available as too is full mobility awareness. The reason for the various types of applications is to allow pre-existing software to run over Chameleon with ease. A third-party application can become a type *C* application with very little extra effort. Extra effort will then result in a type *C* application becoming a type *B* application and so on.

Multimedia is composed of varying types such as audio, video, text, control information. Within these types, exist a multitude of formats such as PCM, JPEG, and MPEG etc. Take the example of a conference application, where control information and files need to be transmitted alongside audio and video. The control information such as who has floor control and files need reliable transport guarantees, whereas the audio and video may be transmitted with a differing QoS. Traditional transport protocols transport the media types through the same stack. If a video stream is filtered through the same stack as an audio stream, the video data will have to adopt the packet size allocated to the audio stream.

Audio in general runs more efficiently with smaller packet sizes [Modiano99]. Isochronous multimedia traffic can tolerate some loss but data that misses its expected delivery time is of no use. Therefore it is more efficient to lose smaller packets than larger packets but smaller packets demand increased header processing in routers. Small packet sizes are not optimal for video data due to the increased size of the media involved. Using an identical protocol stack to cater for all these transport types is not an ideal scenario therefore a more efficient method would construct optimised protocol stacks for each of the media e.g. audio, text and video. Maximum benefit would be achieved if this could be implemented at run-time to cater for the applications particular preferences. A traditional stack belonging to a multimedia application, for example, would send the audio and video in packets of identical size. Research shows that optimal audio packets are smaller in size than video packets [ElGebaly98, SCTE00].

Multiple multicast multimedia groups provide a finer granularity of control compared to using a single video/audio/text stream, because a receiver may subscribe to one or more groups depending on its capabilities (with each multicast group containing varying qualities of media from low to higher resolutions). If a receiver experiences packet loss as a result of network congestion, moving to a lower quality multicast group will reduce congestion, and hence will reduce potential packet loss. This is known as Primary Quality Transformation (PQT). This technique allows media to be composed into broad bandwidth encoded qualities thus all a system needs to do to increase or decrease quality is to move between multicast groups. The Secondary Quality Transformation (SQT) technique compliments this technique by providing fine-grained control of quality within each multicast group by the insertion of transformations in the stream such as compression. The application of multiple multicast group streaming techniques to mobile devices allows the allocation of resources based on local specifications and priorities (see Figure 0-20). Multicast group streaming enables receivers to change the quality of the stream they receive, independently of one another without the source being aware of the change. Considering the feedback problems of multicast, this is a useful property and fits well with an open-loop approach to congestion control of high-speed networks, as when network congestion arises, it is possible to move between quality groups without interruption in service. Service quality should only be slightly reduced but this technique can be highly effective as a last resort for congestion control.

Priorities can be assigned to each multicast group to allow streams to be protected against competing streams. This is an application level QoS scheme and can be implemented easily in Chameleon as all streams pass through a proxy. Pre-set priority levels overcome many problems associated with streaming over wireless links. Long lasting error bursts can severely impact upon applications, causing video frames to be dropped, thus effectively lowering the perceived quality.

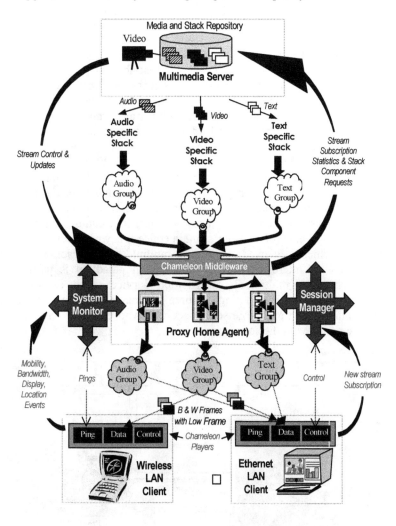

Figure 0-20 : Multiple Multicast Multimedia Groups

Phenomena which interfere with transmissions over wireless channels, ultimately introducing bit errors include atmospheric conditions, physical obstacles and electromagnetic interference. Chameleon supports the seamless operation of real-time streams over wireless links by assigning a priority and a portion of the link's resources, which are protected from being used by lower priority streams. As Chameleon is an open-loop system, a segment with its size defined by the application, is an independent piece of information, similar to the Application Data Unit concept first described in [Clark90]. It is expected that for many multimedia applications, guarantees of reliable delivery will not be necessary for various media types, and some segments could be dropped at times of heavy congestion. In addition, some of these applications may actually be quite tolerant of delays, as described in [Clark98, Tan01]. Particularly for lower priority components, applications would be expected to recover gracefully from loss of segments, or adapt to changes in the delays of their arrivals. Performing transformations on multiple streams is suited to the approach of a source transmitting multiple coded media streams from which the receivers pick according to their individual specifications and capabilities. The benefit of this approach is that there is reduced complexity due to the absence of feedback control mechanisms, which are often redundant for continuous media. Here the source's main concern is to deliver various media streams onto a multicast channel, with no emphasis on where they end up and how they are used. A clients (or receiver's) main concern is what to extract from a channel, which is viewed as offering multiple streams, some or all of which are of interest. This communication paradigm is particularly appropriate for multimedia distribution services such as video-on-demand systems where a single source generates video (and associated audio) distributed to a large set of receivers who generally have little or no interaction with the source.

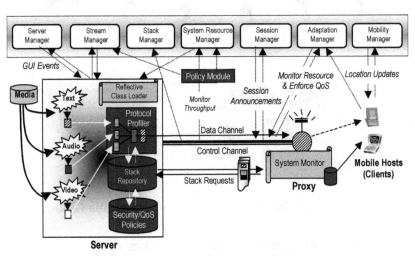

Figure 0-21 : Architectural overview of Chameleon

Chameleon addresses application, application control, and transport layers. The application layer consists of the multimedia application (e.g., a video-on-demand application) which is responsible for retrieving the stored audio/video file with captions/subtitles (multilingual), composition at the sending end, and the audio/video client which is responsible for decoding and displaying the video frames at the receiving end. Application control consists of a media filter at the sending side to demultiplex each stream into several sub streams, and media filter at the receiving end to multiplex back one or more sub streams for the audio/video client. These multiplexed streams (transport layer) differ from common practice in that these streams are not logically grouped together and shipped over the wire. Instead, the media elements are divided into audio/video/text by the event filter and distributed to separate groups in accordance with application layered framing practice and then the receiving filter directs the streams to the relevant media application, thus the streams retain their distinctiveness. Media may be stored in separate files on the server and so that there is no need to split the media in real-time. The application media filter receives events from the application which may categorised them as text, audio or video. A session manager is consulted to see how many groupings of each category are required. The normal is one for text, and three each for audio and video. The text stack is composed as a reliable stack. The audio/video stacks are both UDP differing in default packet size and header sizes. Each media is sent to separate multicast groups where the well-known addresses are obtained from the session manager. Each of the three sub-groups of audio and video will require a separate multicast address. Since the network load changes during a session, a receiver may decide to join or leave a multicast group, thereby extending or shrinking the multicast trees.

Channels are a multicast medium into which sender applications basically 'push' streams, and to which receivers can subscribe to receive those streams. A channel maps into an IP multicast group or a point-to-point UDP connection with Uniform resource locators for naming purposes such as *Chameleon://227.34.3.63/lectureHCI/HQ_Audio*. With increasing scale, dissemination actually saves bandwidth because it eliminates the flood of duplicate requests and responses when multiple clients all request the same stream. Multicast communication allows applications to be relocated from one machine to another and to distribute data from one sender to many receivers efficiently also catering for fast and slow receivers. Publishers (i.e. Senders) and subscribers (i.e. Receivers) must conform to a public interface. A side effect of this publish-subscribe paradigm is that applications written in other formats such as Microsoft's Common Object Model (COM) and JavaScript can be written to adhere to this protocol or wrappers can be written over existing applications to work seamlessly with Chameleon. The active network proposals discussed earlier, target network programmability without being content-aware. Chameleon, in contrast targets content-aware application-level programmability. The rapid increase in media types necessitates a network infrastructure that allows clients and servers to be free from media dependency and burdens of managing content & client heterogeneity. This can be extremely important for streaming media because of its demanding resource requirements for processing, translation, and transmission thus middleware must combine media awareness with a high degree of intelligent adaptivity in order to truly serve heterogeneous clients.

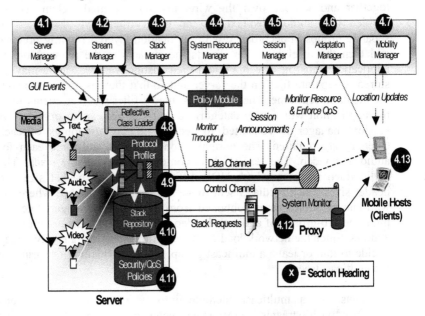

Figure 0-22: Section headings for framework discussion

The following sections examine the components that compose Chameleon. Figure 0-22 illustrates these components and the book section headings, which provide a more detailed explanation of their function. The architecture is examined in more detail by tracing the components that are involved in streaming media from the server to each of the mobile clients.

4.1   Streaming Server & Multicast Group Management

A media-streaming Java server delivers multiple multicast media streams to clients.   The industry-standard Real-Time Protocol/Real-Time Streaming Protocol was used for default web casts. This ensures that no file is ever downloaded to a client's hard drive, which is essential if targetting resource constrained mobile devices. Thus, media is played, but not stored by the client software as it is delivered. The server can transmit through IP multicast or unicast depending on whether the URL denotes a multicast address or not. Multicast is recommended in order to achieve the full functionality of Chameleon and the performance benefits (including bandwidth savings). Each video stream can be sent through a live web cam feed using a *Capture* component or else a pre-recorded media sample is retrieved from a series of pre-stored media clips on the servers drive. In the case of a live capture, the media is then filtered though an encoder for compression. The compressed data is returned to *RTP.Processing*, which forwards it to the framing component for packetisation. The *Transmission Application* component delivers it to the network layer for transmission.

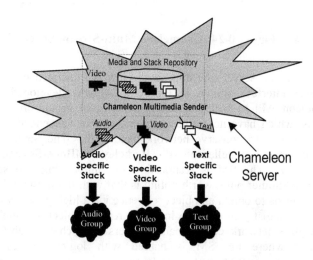

Figure 0-23: Snapshot of Streaming Server in architecture

Media files have certain attributes such as encoding type, frame size, frame rate, and the data type and these have an effect on the quality and size of the file created thus Chameleon allows new content developers to expand the capabilities of the framework to handle MPEG-4, QuickTime movies with Sorenson Video, and even user-added QuickTime components like On2's open-source VP3 or Apple's optional MPEG-2 component [Quicktime03]. This should provide the video quality required for a commercial application.

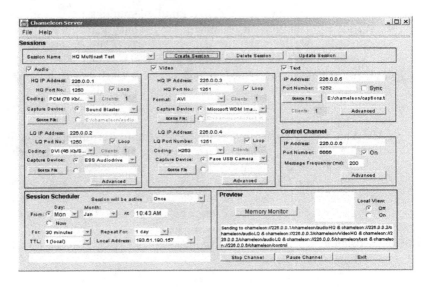

Figure 0-24: Chameleon Multi-Stream Server

A multi-stream server (the high level view is illustrated in Figure 0-23 and user interface shown in Figure 0-24) was developed using the Chameleon API to facilitate multiple multicast group streaming. Any streams, which have no active receivers, will be suspended after a pre-specified time to conserve network bandwidth. The Stream Manager (SM) is loosely modelled on *sd* – Berkeley's MBone Session manager [Handley97]. The SM is responsible for announcing and scheduling sessions. Another point worth noting is that with traditional middleware, a client needs to obtain an object reference via a highly available naming service or registry in order to invoke a remote object. Chameleon does not require network centric name servers as a channel abstraction is provided where the SM is named with logical names such as `chameleon://224.2.1.42:4444/sm/testconf/`. Clients can then join & leave sessions dynamically without having to start the system anew. A name can be given to a session as illustrated by the text box in the top left of the user interface in Figure 0-24. The buttons to the right of this allow a session to be created, deleted and updated (should changes have been made to the Multicast IP group addresses).

The main section of the GUI is taken up with sections for streaming audio, video and text. These allow the entry of IP Multicast addresses for the High Quality and Low Quality audio and video streams alongside designated Ports. The coding for the stream can be chosen from a pre-defined selection and the capture device can be specified as a live source or a pre-encoded file. The text part of the GUI simply allows the entry of the Multicast IP address with port and the source file to use for the ASCII text. The Synchronisation button allows text such as sub-titles to be displayed synchronously with audio and video. This is achieved by placing time delimiters in front of lines of text in the source file.

The user interface for the Session Manager can be seen in the bottom left corner of the stream manager as illustrated in Figure 0-24. This allows a session to be run once, daily, weekly or monthly. It also allows setting the time for the session to start and duration. The Time to Live (TTL) can be set to broadcast to the local machine, local network, UK, Europe or the World. A preview window can be seen in the bottom right of the User Interface and this enables local viewing on the server of the broadcast stream. A memory monitor button is placed alongside which displays free and allocated memory on the server. This was used extensively in testing in order to verify the load on the server during trial broadcasts.

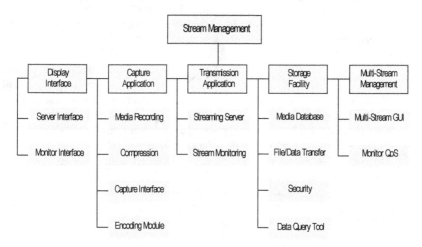

Figure 0-25: Stream Management Classes

A partial view of the stream management classes is displayed in Figure 0-25.

Display Interface. The display interface shows the various streams and the qualities of service that they are capable of providing. It also allows the administrator to set and control video and audio content such as the codec to be used and specify sources such as recorded media or live web casts.

Capture Application. The capture application is capable of recording voice and video data from a microphone and web cam respectively. The capture application also interfaces with the data compression modules which transcodes the stream into a format such as PCM.

Transmission Application. The Transmission application is responsible for streaming stored media files from the database (or live web cast). The transmission application module is capable of working with RTP [RFC1890]/RTSP [RFC2326] or plain HTTP multicast broadcasts.

Storage Facility. The storage facility stores the media clips. The storage facilities other main task is to allow the streaming manager to retrieve the correct file at a later date.

Multi-stream Management. The Stream Management component interacts with the System Monitor when Quality of Service Multicast group changes occur due to adaptation decisions made elsewhere. Therefore the media quality of each stream is upgraded or downgraded accordingly.

With application controlled QoS (as in Chameleon), different applications running on the same system may have very different adaptive behaviour when QoS variations occur. Some of these may consume a considerable amount of system resources to perform their desired adaptive behaviour, (provided resources are available) while others may not perform any adaptation at all. Clients subscribe to multicast groups to receive media packets. Chameleon extracts the different media types from within the application and creates suitable run-time protocol stacks to enable streamlined transport communication to be invoked. The Chameleon middleware is responsible for re-assembling the flows at the receiver.

4.2   Stream Management

The sequence of events in starting a Chameleon session is illustrated in Figure 0-26. The server can be started first. Streams are broadcast to random multicast groups. The proxy is then started and retrieves the list of active streams from the server. The proxy then subscribes to the groups and rebroadcasts the streams to other multicast addresses. Stack profiles are referenced at this stage and streams needing reconfiguration according to the profiles are transformed accordingly. The mobile is started and subscribes to the well known chameleon multicast control channel and subscribes in accordance with desired quality. At this stage PQT and SQT techniques (see 4.6) commence their work.

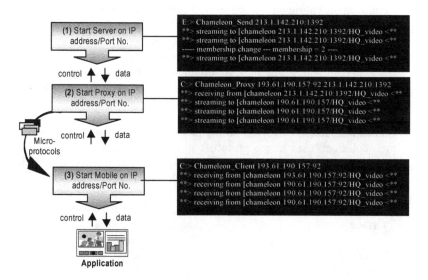

Figure 0-26: Startup Sequence of Chameleon Components

Chameleon is concerned with the transport of media elements to the end-user which may differ over time in that a user may choose to retrieve audio only, or sound and vision or sound, audio and sub-titles. When dealing with multimedia, which is sent through separate transport stacks, a synchronisation facility, which adapts to the users decisions, must be in place, so as to allow presentation of the media in a co-ordinated manner. One rule was to make the audio the master player so that video and sub-titles were synchronised to the audio playout. This it was felt was the most natural means of keeping the media in synchrony. The Media Framework [JMF01] was used for the synchronisation of the various media files. In synchronising the audio and the video, a video interframe time is calculated as a function of the total number of video frames and the total time of the audio portion of the uncompressed clip. The synchronisation of sub-titles with audio/video is achieved by storing the playout times of the sub-titles in a text file that is stored on the server.

4.3   Stack Management

Central to providing an adaptable QoS is the ability to maintain multiple protocol stacks. A protocol stack consists of a linear list of protocol objects and represents a quality of service such as reliable delivery or encrypted communication. The framework provides the services necessary for supporting new communication protocols and qualities of service. Chameleon consists of a set of Java classes for representing Uniform Resource Locators, protocol stacks, the framework API and media packet objects similar to Horus [Renesse96, Maffeis02]. Chameleon stacks have the ability to be configured at run-time. The protocol stack uses micro-protocols in its implementation where each micro-protocol (or layer) enforces a part of the quality of service property guaranteed by the protocol stack as a whole. Creating a layer for each property and stacking them on top of each other achieve the properties desired by the user of a stack as each layer contains the same interface.

This research proposes that synthesis of 'fine grain' protocol functions should replace the coarse grain protocol design of traditional protocols (e.g. TCP Reno). It can be argued that the integration of all the application communication requirements (including transmission control, synchronisation and presentation encoding) in a single optimised protocol graph will result in increased performance, which is in line with the ALF architecture. The protocol profiler is used to configure a protocol that satisfies the optimal protocol configuration as defined in the profiler for each media. Protocol functions such as acknowledgements, flow control and check summing are located within the Chameleon source code 'tree' as separate classes. Given the timeout and retransmission mechanisms of reliable transport protocols, each class is multithreaded with each protocol configuration being a protocol graph, which defines a set of stack elements and their relations. Stack elements are implemented in classes with each class encapsulating a typical task such as error control, flow control, encryption or decryption. Normally, there are several classes available for a single protocol function (e.g., a system may implement a FIFO or a LIFO queuing algorithm in the end-system buffer).

Stack Description	Class Name	Layer	Requires
IP Multicast	IPMCAST	1	-

Acknowledgements	NACK	2	IPMCAST
Queuing	FIFO	3	IPMCAST, NAK
Queuing	RR	3	IPMCAST, NAK
Fragmentation	FRAG	4	IPMCAST, NAK, FIFO
Cryptography	CRYPT	5	IPMCAST, FRAG, NAK, FIFO

Figure 0-27: Stack Profiles

Stacking of layers randomly may not make semantic sense in many cases thus Figure 0-27 illustrates stack profiles where the Layer column specifies position in stack and the Requires column indicating the mandatory underlying stack components. All layers in Chameleon implement the layer interface, which means that a developer will only need to see the interface class. In order for a client to be able to receive, decompress and/or decrypt data, it may need to download the appropriate layer(s) from a central repository. These layers are downloaded through Java serialisation where it can be cast (on the client) using the layer interface and appropriate methods called thus removing the need for client software to be bundled with all layer classes. This ensures that clients can use protocol functions, which were not available at the time the client software was published.

A message sent by *ProtocolStack* is passed to the protocol stack, which in turn forwards it to the top-most layer. All layers perform some computation and pass the message on to a layer below with the bottom-most layer placing the message on the network. In the opposite direction, the bottom-most layer of a protocol stack will receive a message from the network and pass it on to the next layer above where this layer performs some computation and passes it to the layer above it. The message will be removed when *Stack.Receive* is called. Layers can add protocol specific data, such as a checksum or a key for encryption at any stage. The design decision to use java interfaces allows third party programmers to call *addlayer()* to append a new layer object or indeed a *setlayers(arraylist)* which allows the passing of an array of objects which implement the interface. The Protocol Profiler according to a properties argument defined when creating an instance of Stack creates protocol layers. A properties string might be:

```
UDP(mcastaddress=192.4.3.5):NAK:FRAG(size=8096):FIFO:ENCRYPT
```

The Protocol Profiler parses the string, creates the corresponding instances ("UDP", "NAK", "FRAG", "FIFO" and "ENCRYPT"), sets their initial data ("mcastaddress=192.4.3.5", "size=8096") and connects them one to another and the top layer is connected to the protocol stack

98

object. Then the Protocol Profiler iterates through the protocol stack and starts each layer in turn. When shutting down a stack, the Protocol Profiler stops each layer in turn, giving it time to process outstanding messages and then destroys the stack. The diagram in Figure 0-28 depicts three protocol stacks. The text stack is a stack that addresses applications that need to communicate reliably through private channels, CRYPT takes charge of encrypting and decrypting media packets on the fly while NAK takes care of retransmitting lost media packets and of flow control. The stack is composed as follows:

```
Stack text = new stack ("CRYPT:NAK:MCAST");
```

The following video protocol stack could be used by applications that require that media packets be communicated as efficiently as possible as large media packets not fitting the UDP datagram are fragmented and reassembled by the "FRAG" layer. This stack is composed as follows:

```
Stack video = new stack (''FRAG(size=2048):MCAST'');
```

Steps involved in a stack such as "FRAG:NAK:MCAST" may involve a media packet being fragmented into fixed-size datagrams by that FRAG object that may be sent through an IP multicast socket. FRAG passes the datagrams down to the NAK (negative acknowledgements) object where NAK caches each datagram in case it is lost on the network (and receiver requests its transmission). Finally, MCAST submits the media packet over an IP multicast communication channel. The connection end-point abstractions that Chameleon uses are implemented using lightweight stack objects layered above connectionless protocol objects such as UDP/IP. ALF states that diverse application requirements are to be met by leaving as much functionality as possible to applications. The middleware is designed to meet only the minimal definition of multicast and if the need arises, machinery to enforce a particular delivery order can be easily added on top of this delivery service. Distinct media formats deserve distinct transportation treatment. Proxies filter the data depending on the source data stream e.g. audio (Microphone), Video (Camcorder) or text (File transfer) and composes one of a library of protocol stacks suitable for transmission of the media as illustrated in Figure 0-28.

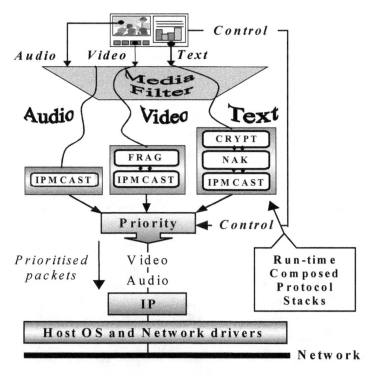

Figure 0-28: Proxy filtering of isochronous media streams

The result is that separate streams from the same application are multicast to the same IP group address, and filters recompose the streams into an integrated application. Protocol stacks can be compiled as late as run-time depending on the need for re-adaptability. There is a noticeable cost for stack reconfiguration but it is a reasonable overhead if it can be amortised over multiple subsequent data exchanges. Standard recognised objects are audio, video or text objects. These can be expanded to include more objects and specialised media types within these groups. All components in the stack implement the *Chameleon.SenderListener* Interface and *Chameleon.SenderMulticast* Interface. Each stack component then has the chance to perform processing on the data such as filtering, parsing etc. The Java ClassLoader class was extended to allow loading of new stack components from a remote repository (discussed later in section 4.9.3). This allows the dynamic downloading of a new protocol component from a WWW server, FTP server or local device when needed. This allows Chameleon to posses an extremely lightweight library on mobile devices.

4.3.1   Garbage Collection

An application begins execution with a pre-defined set of stack
elements. In the course of execution, elements of the stack may be
inserted into the stack by being read from disk or downloaded across the
network. In addition elements may also be removed. There is a design
issue involved here as to when to reclaim memory from the heap (from
floating stack elements) as illustrated in Figure 0-29.

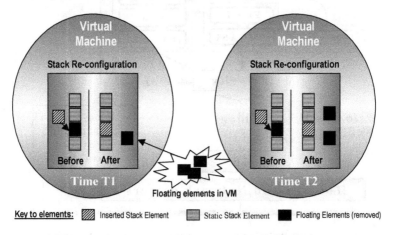

Figure 0-29: Floating Stack Elements within Java Virtual Machine

The approach taken here is one of allowing Java's mark-and-sweep
algorithm to recycle memory cells once the free memory goes below a
certain threshold. This has the obvious benefit of relieving the
programmer from deleting objects and eliminating memory leaks. There
is an overhead involved with this depending on how often the stack
elements are swapped, but a future design goal is to attempt to use this
information to facilitate defragmentation and compaction of heap
memory

4.4   System Resource Management

The heart of the adaptation decision-making and System Resource
Management is the System Monitor. The monitor has been created in a
manner, which allows a number of existing (and future) components to
'plug' into. The System Monitor exists on the client in order to monitor
quality of service and resource capabilities. A SystemMonitor class

contains current system information, which reflects QoS parameters of the client, such as processor load, available bandwidth, screen characteristics and more. Each client in order to determine preferable formats for reception of media instantiates a SystemMonitor() class. For example, a client with low processor speed and high processor load might not be able to receive MPEG streams. The code in Figure 0-30 shows the declaration of SystemMonitor class.

```
public class SystemMonitor implements Serializable {
        int processorSpeed = 0;        // in MHz
        int processorLoad = 0;         // in %
        int estimatedBandwidth = 0;    // in bps
        int availableBandwidth = 0;    // in bps
        int totalMemory = 0;           // in MB
        int availableMemory = 0;       // in MB
        int screenWidth = 0;           // in pixels
        int screenHeight = 0;          // in pixels
        int colourDepth = 0;           // in bits
        int   numberOfSpeakers   =   0;                    //
        mono/stereo/surround
}
```

Figure 0-30: System Monitor QoS parameters

Four components have been developed and are present in Chameleon to aid decision-making policies from a number of aspects. A discussion of each monitor (see Figure 0-31) follows.

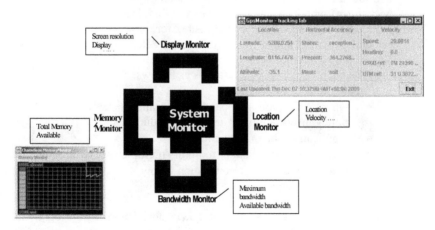

Figure 0-31: Plugging in Monitoring Components to Monitor API

102

Bandwidth Monitor. An adaptive system needs to be capable of monitoring raw data throughput in order to respond to network variability. Performance measures are generally application dependent, but will typically be a function of responsiveness and data delivery rate. Hence Chameleon provides the hooks for a monitor stack element to be inserted tailored to an application which has in place means to monitor the data and manage the bandwidth usage of the system when the data requirements of the various media exceed the available bandwidth. Statistics, which can be derived using these classes, are listed in Table 0-5.

• *arrivals this flow (pkts)*	• *arrivals this flow (bytes)*	• *early drops this flow (pkts)*
• *early drops this flow (bytes)*	• *all arrivals (pkts)*	• *all arrivals (bytes)*
• *total early drops (pkts)*	• *total early drops (bytes)*	• *total drops (pkts)*
• *total drops (bytes)*	• *drops this flow (pkts)*	• *drops this flow (bytes)*

Table 0-5: Stream Statistics

Late arrivals of packets may be an indication of some bottleneck in the system, in which case the target can inform the origin about the overload and cause it to scale down the stream. Once the overload situation has passed, the stream may be scaled up again. In the scaling implementation of Chameleon, individual streams are mapped onto multicast groups, each with its own set of QoS parameters. The transmission quality can then be adjusted either with fine granularity within a group (SQT mechanism) or with coarse granularity by adding and removing groups (PQT mechanism). The prerequisite for scaling mechanisms are functions, which allows the system to detect network congestion. Chameleon achieves this by monitoring end-to-end delay and packet loss rate. The bandwidth manager's task is as follows.

- On initialisation the bandwidth manager retrieves bandwidth policy information for a specific media stream.
- Applications begin transmitting.
- The bandwidth manager monitors multiple stream throughputs.
- Upon packet loss or congestion below a threshold – the system monitor is called.

When a packet arrives, a monitor queue object notifies the bandwidth manager object of this event. The bandwidth manager using this information monitors the queue as illustrated in Figure 0-32. A packet has an expected arrival time and a packet arriving later than expected indicates congestion. Chameleon calculates the expected arrival time as the 'logical arrival time' of the previous packet plus the stream period. The logical arrival time is the arrival time observed when bursts are smoothed out, that is, when early packets are artificially delayed such that the stream rate is not exceeded.

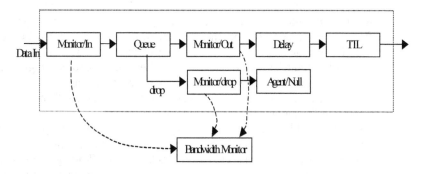

Figure 0-32: Monitoring Queue

The Bandwidth Manager works by monitoring the UDP-level latency on each link by sending a 32-byte UDP packet to active participants. If an echo is not received within the maximum ping period or four times the current round trip time estimate, the packet is retransmitted. Using UDP can sometimes lead to loss of messages but this should only lead to a short-term loss of efficiency. A problem arising when only monitoring end-to-end delay is the definition of a threshold value for congestion detection. Chameleon detects congestion through the packet loss rate but the lateness of a single packet should not immediately trigger PQT, as the congestion may be short. If a sequence of packets is late (or packets dropped due to buffer overflow), it is assumed that the network is congested thus invoking the PQT mechanism to move to a lower quality group. The monitor provides a container for storing byte counts of data along with corresponding start and stop times for recording the time interval for stream sessions. The monitor stack element also provides an *addSample()* method for adding bandwidth measurement where samples can be queried for statistics about the stream such as *getAverageRate()*, *getRateFor()* and *getLastRate()*. An attempt to minimise the effects of the Hiesenberg Uncertainty Principle [Cassidy92] is done by reading and processing large amounts of data in each cycle with relatively few data rate measurements. This poses the risk of limiting our ability to monitor throughput variations over time thus relying instead on rate averages over more significant time periods. In our case, the effective raw data rate for the system over one read/decode/write cycle is the total amount read (d_T), divided by the sum of the times for the reading (t_r), decoding (t_d), writing (t_w) and monitoring (t_m) of the data.

104

$$R_t = \frac{d_T}{t_r + t_d + t_w + t_m}$$

A bandwidth manager is crucial for mobile environments as streaming media to mobile hosts when competing with TCP traffic is severely affected [Widmer2000]. As the network does not distinguish between normal data packets and mobile IP control packets - beacon messages, and binding requests are frequently lost when flows send at a higher rate than the available bandwidth. This can prevent the mobile host from performing a handoff to the next base station. A bandwidth manager is imperative for an acceptable throughput distribution of competing flows where bandwidth is scarce thus it is argued that mobile IP control packets should be given priority over normal data packets. The bandwidth manager can prevent restricted flow packets from flowing through the protocol stack effectively preventing packet collisions caused by rogue flows. Applying priorities to control packets can be achieved by using two separate queues for data and control packets with a higher priority being assigned to the control packet queue.

Location Monitor. This component allows mobile clients to roam and take advantage of foreign base station streams. The mobility of a computer user suggests a new class of applications called context-aware computing. These applications are made aware of the context in which they are run, based on a limited amount of information covering a user's proximate environment, to exploit the constantly changing environment. There are three important aspects of context: where one is, who is around, and what resources are nearby. Being aware of it may help promote and mediate one's interactions with devices, computers, and others, and help navigate unfamiliar places or find devices such as the nearest server [Terry94, Petrovski03, and Barnes03]. The nearest server may cease to be the nearest due to migration as a physical distance may not correspond to a network distance (e.g., when crossing administrative domains), the communication path may grow out of proportion with respect to actual movement. A longer communication path not only consumes more network capacity but also has more intermediaries, and thus, longer latency and greater risk of being disconnected. A context-aware application can avoid such problem by dynamically transferring service connections to closer servers [Nguyen01]. The combination of Mobile IP and Wireless 802.11 are sufficient detecting movement in and around base stations. It must be recognised that future systems are likely to utilise location information in adaptation decisions therefore to cater for this, Chameleon possesses a GPS Location Monitor class that interfaces with a GPS receiver using the NMEA 0183 version 2 protocol. The component stores the following settings: Current latitude, Current longitude, Current altitude, Current speed, Current heading, Mean horizontal accuracy, Present horizontal accuracy, Time at which the position was last calculated and Current status of the receiver. The component is accurate to within 3 meters. At present, the actual co-ordinates of each proxy locations are stored in a proxy-coordinates.txt file as in Figure 0-33 although recognise the limitations of such a system.

Base station ID	Latitude	Longitude	Users	Size
Magee.A.2	5422.0233	0232.3444	All	200
Magee.B.3	5733.0445	0189.3434	Super	150
Magee.B.1	5533.6445	0198.3433	Super	300
Magee.A.3	5203.5433	0299.6743	All	300

Figure 0-33: Base Station GPS Co-ordinates

To enable access to serial communications lines, the GPS driver uses the javax.comm library that is available as an extension to the Java Development Toolkit. The design of the NMEA message parser uses the ability to reflect on Java classes (java.lang.reflect), which allowed Chameleon to enumerate the members and types of classes. Using this feature, the syntax of NMEA messages could be specified, which is an advantage, as GPS manufacturers tend to add proprietary NMEA messages to enhance functionality. GPS allows us to mark the location of base stations. This information can be used in conjunction with mobile prediction algorithms to 'guess' the direction, which a mobile device is moving (as people tend to follow a straight line). Upon detection of the movement of the mobile, techniques such as requesting a stream from the server and caching it can be used. This means that once the mobile enters the 'cell' where the data has been cached, the hand-off process should be much smoother as the mobiles home agent has already begun the streaming process thus delay is minimised. It is hoped to incorporate mobile prediction algorithms in a future version.

Display Monitor. Devices vary in their display capabilities. This component detects a devices screen resolution thus affecting the parameters of any delivered streams. The screen resolution can be discovered via the java.awt.Toolkit class. This information is then passed to the SystemMonitor class to allow further QoS decision-making. Sample code is illustrated in Figure 0-34.

```
public class ScreenRes {
        public static void main() {
                Dimension screenSize = Toolkit.getDefaultToolkit().getScreenSize();
                String reconfigfile_Settings = null;
                URL url = null;
```

106

```
switch(screenSize.width) {
        case 320: ..SCREEN_VLOW = true; ... // 320x240
        break;
        case 640: SCREEN_LOW = true; ... // 640x480
        break;
        case 800: SCREEN_MED = true; ... // 800x600
        break;
        case 1024: SCREEN_HIGH = true; ... // 1024x768
        break;
        case 1280: SCREEN_VHIGH = true; ... // 1280x1024
        break;
        default:  SCREEN_UNDETERMINED = true;
        break;
}
```

Figure 0-34: Screen Resolution Detection Code

This can later be used to present various mobile devices with pre-designed optimal interfaces. To date Chameleon has only attempted to provide two separate client player interfaces dedicated to two screen resolutions groups. There is a standard interface that maps nicely onto resolutions from 800x600 to 1280x1024 (i.e. laptops, desktops etc) and another restricted GUI that is tailored for displaying on screen resolutions of 320x240 (i.e. PDA's and sub-notebooks).

Memory Monitor. The Memory Monitor component tracks free memory and triggers warning events upon entering regions defined in the memory_threashold settings. Detecting total and free memory is done in Java. The memory monitor component depicted in Figure 0-31 like all other system monitors in designed to be easily extendable. The process by which the monitor collects information is called capturing. By default, a memory monitor gathers statistics on allocated and free memory and alarm thresholds can be built into the protocol profiler, which controls reconfiguration of protocol stacks so that events can be triggered as a response. In parallel with capturing information, a memory monitor window displays available memory. Once data has been captured, the data can be saved to a text or a capture file, and can be opened and examined at a later time. This can prove useful when debugging new protocol objects. One usage of the memory monitor is that of a device receiving only an audio stream yet is also low on memory. In this case, any attempts to add a video feed (or upgrade to Higher Quality streams) should be prevented until memory is freed.

4.5   Session Management

A *MediaSource* object packetises the media elements using a *ProtocolDecode* object to encapsulate the data in RTP protocol data units and forwards the packet stream by means of a *Connection* object. The management of an application is in the main controlled by a *HomeAgent*, *UserInterfaceManager*, *StreamManager*, *SessionAgent*, *SessionManager*, *BandwidthManager* and *MobilityManager* (as shown in Figure 0-21). The *UserInterfaceMananger* provides GUI objects for configuring and monitoring architecture components. The *StreamManager* administrates and assumes responsibility for registration of all the data streams. The *SessionManager* is responsible for Sessions, Senders, Clients and Session Descriptions while *SessionAgent* imports and exports information concerning local and remote sessions using session description-announcement protocols. The *HomeAgent* is the central repository for information on mobile client locations.

An active session comprises *Channel objects*, which carry media streams and control information, a *ProtocolProfiler*, which encapsulates media elements in protocol data units, and a *Profile*, which is used to configure the *Stacks*. A *Session* object interacts with the underlying system's network QoS control entities, by means of the *BandwidthManager* object that is responsible for negotiating and maintaining network resource reservations for flows. The *Profile* object also specifies the maximum packet rate and the bandwidth share of a session. The *SessionManager* is responsible for *session termination* and *cleanup* where all threads of control are terminated. Protocol clean-up actions must also be triggered, such as delivery of RTCP bye packets, or leaving multicast groups.

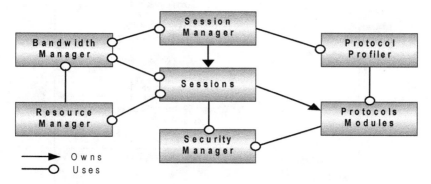

Figure 0-35: Chameleon Components

This model (illustrated in
Figure 0-35 and Figure 0-36) assumes that all users can access a common multicast address hosting the session manager service. This service advertises bindings of multicast addresses to particular media flows on a

well-known bootstrap address according to the Session Announcement Protocol (SAP) [Handley96, Zhao02] where a multicast scope mechanism constrains the distance multicast datagrams may propagate in order to enhance scalability. The Session Manager contains an ASCII text based protocol for describing multimedia sessions and their related scheduling information so as to convey information about media streams in multimedia sessions to allow the recipients of a session description to participate in the session. The session manager separates media control from media transport. A reusable session manager, which is separate from the user interface or particular applications helps avoid duplication of effort. This separation promotes the development of replaceable agents to cater to diverse hardware capabilities and user preferences.

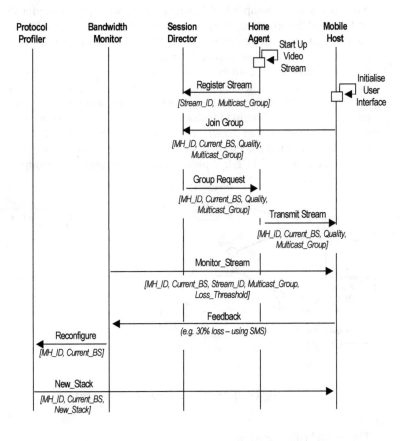

Figure 0-36 : Interaction Time-Line

Participants share a common control channel (Figure 0-37) that enables

109

coordination and communication of relevant events and information. Transcoding and rate-control is controlled by commands issued over the control channel. From a control point of view, the transcoder is a passive entity, which exports an interface and is controlled by commands on the control channel, which translate application instructions into control channel commands, which then appropriately configure each transcoder.

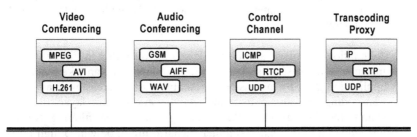

Figure 0-37 : Conference environment using control

The traffic over the control channel may be one-way or two-way where applications announce without expecting replies or applications announce expecting a response. Chameleon's native control and streaming transport protocol is RTP compatible that provides interoperability with existing Internet media tools. The RTP control protocol (RTCP) provides feedback on the quality of the data throughput, which is important for diagnosing failures, monitoring performance, and ultimately, for dynamic adaptation to congestion. Flow statistics are distributed from active sources via sender reports (SR) and from receivers back to the entire session via receiver reports (RR) which include metrics such as cumulative packet count, cumulative byte count, cumulative count of lost packets, short-term loss indicator, and an estimate of the jitter in data packets.

The fixed negotiation schemes discussed in section 1.5 are not appropriate for complex multimedia requirements therefore a design consideration of this framework is to provide a central resource allocation/synchronisation unit, which controls the adaptation behaviour of each application and balances the resources required during the adaptation within the client. Localisation of control offers a number of advantages over end-to-end control where the propagation delay between client and flow control process may be much smaller, thus allowing more accurate and reactive feedback control. The signalling messages between client and control nodes traverse fewer hops and are therefore less susceptible to suffering from the effects of a congested network.

4.6   Adaptation Management

Modern distributed applications, such as enterprise computing and mobile multimedia applications encounter unpredictable environments due to user mobility and varying resource availability. To adapt, systems must identify the need for a change, decide on the change and implement it in a timely manner. This section deals with adaptation within Chameleon.

Adaptation can occur at multiple levels as illustrated in Figure 0-38 For instance in the case of streaming media, adaptation can take place in the application layer by increasing the compression, decreasing image size or transcoding the stream to mono. At the middleware layer, the server source for the stream could be changed or frame filtering could be introduced into the path. Another scenario might be where high bit rate errors are encountered over wireless links and some sort of forward error correcting module is injected into the path.

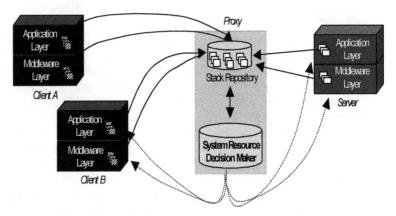

Figure 0-38: Multi-Layer Adaptive Architecture

Chameleon adopts a centralised adaptation architecture where system monitor components are embedded in the application components and/or middleware. A centralised System Resource Decision making component periodically receives event information from these monitors and reacts according to QoS policies defined by the system administrator. The adaptation process repeats a cycle of estimating, deciding and acting with the use of observation variables, which capture relevant aspects of the system status. One of the more commonly used observation variable is throughput. Monitors are components that aid the Decision Making Component in implementing the decision to restore the system to a desirable state of operation. These components may be built into the application layer or the middleware layer as illustrated in Figure 0-38.

Due to the nature of next generation communication networks using different kind of wireless access and added mobility, applications will have to react rapidly to variations in resource availability. To cope with temporarily unavailable network resources, multimedia applications have to be elastic in adapting media representations without excessively sacrificing the perceived quality of service. To address both mobility and QoS issues, two alternative but complementary architecture solutions have been identified. The first approach purely leverages existing protocols and components defined (or being defined) by the IETF, and tries to provide the necessary extensions to them. The choice of this organisation is due to the fact the Chameleon architecture is IP-centric and so therefore no modifications of existing applications are needed. The second approach presumes instead the availability of some middleware, which is providing the major functionality for dealing with mobility and QoS issues, as well as offering several Application Programming Interface (API) functionalities for to-be-developed applications. Both viewpoints are therefore synthesised in a modular fashion, indicated by different types of application classes in the architecture.

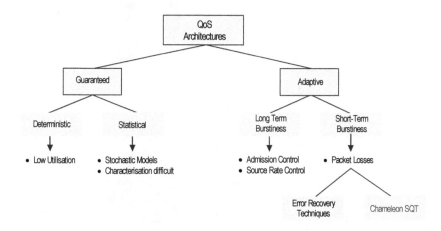

Figure 0-39: Architectural view of SQT in the QoS scheme

There are likely to be lots of variations and developments at the lower levels, and the lower layer protocols will only provide a certain level of QoS that needs to be enhanced for many applications. Hence the need for a middleware layer to provide suitable abstraction from the networking layers and facilitate session layer QoS processing. Mobile terminals moving into regions with low signal quality or handing-off to new access points, may violate the QoS contract with the network, which can cause the frequent dropping of connections. This requires QoS adaptation and even re-negotiation. All these conditions require the applications to be adaptive in a sense that applications have to react to varying resource availability inside the network and the end systems. In order to simplify the programming of mobile broadband applications and to allow for support of dynamic QoS changes, these active adaptation mechanisms should be hidden to application programmers. The idea of shifting adaptation mechanisms from the application level to a flexible middleware featuring QoS functions will thereby result in simplified application development for mobile environments. The goal of Chameleon is to allow any kind of application to get the desired level of support from the system in other open environments like the Internet.

Figure 0-39 illustrates where SQT fits in relation to common approaches to providing network QoS.

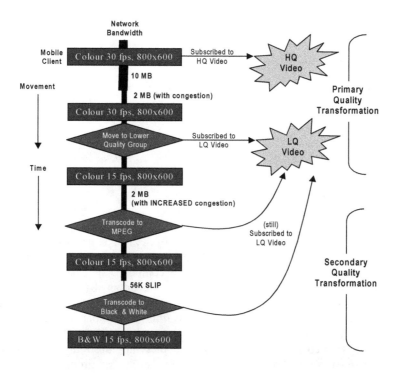

Figure 0-40 : Primary and Secondary Quality Transformation

For instance, the Primary Quality Transformation algorithm assumes responsibility for coarse grain adaptation decisions. This involves moving between multicast groups upon violation of group bandwidth limits. Secondary Quality Transformation assumes responsibility for responding to quality fluctuations within each group. The SQT technique works through the use of transcoder mechanisms, which transform the data as it flows through the proxy. Transformations could include downgrading a full colour 30fps AVI movie to a Black and White 15fps MPEG movie as illustrated in Figure 0-40.

PQT with priority relies on third party traffic being disabled (or rate controlled). This can be achieved through the technique of blocking ports. Traffic types within Chameleon are assigned port numbers to designate media type and these work alongside the well know port numbers assigned to traffic such as FTP, TCP etc. in order to bring about prioritised media traffic streams.

	Chameleon 'Firehedge'	
Video (100k)	Port **6667**	
Audio (32k)	Port **6677**	
Control (4k)	Port **6688**	
FTP (128k)	Port **21**	

Figure 0-41 : Chameleon Port Blocking

Figure 0-41 illustrates the port blocking of the FTP stream on port 21 and the TCP stream on port 80. For the PQT with priority technique to work properly, the end-to-end path must be composed of Chameleon filters enabled in 'firehedge' mode. This title was adopted (i.e. hedges being much weaker than walls) as opposed to firewalls but there is limited functionality in a technique such as this in comparison to industry standard firewalls which block, monitor and classify traffic in a much more detailed manner. Figure 0-42 depicts an illustration of a media stream subscribed to a medium quality multicast group. This group broadcasts media ranging in quality from 400k to 1000k. The group at both the low end and the high end of the scale contains a 100K 'danger' region. Monitoring within the danger area and violations are handled by PQT whilst monitoring with the 400K interim regions is handled by SQT.

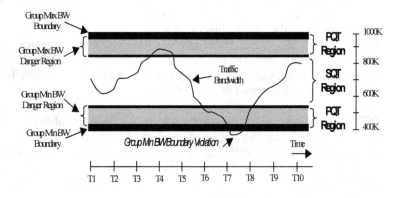

Figure 0-42: Quality Adaptation Domain

The adaptation algorithm is the proxies' responsibility for deciding whether a PQT group move should be made or if an SQT filter should increase or decrease bandwidth. A measurement of the current packet loss rate is recorded using a sliding window of variable length S packets. Take the case of a stream, which is receiving the lowest quality of media (e.g. black and white CIF video over a wireless bandwidth). If the client reconnects to the LAN, the bandwidth manager will record increased packet delivery beyond a certain threshold for a certain period, as specified in low->high util‰ (increase from B&W CIF to Colour), and ave interval‰ (average) parameters. The client may also return to mono video (lower quality) when the data traffic activity yet again has fallen below a certain threshold for a certain period, as specified in high->low util‰(decrease from high to low), and ave interval‰.

Adaptation occurs dynamically on an 'as needed' basis with all threshold points being defined in the Session Manager's stream profiles. Threshold points are flexible and allow the definition of adaptation points such as how long traffic is to remain at a specified percentage level before renegotiating QoS. Figure 0-43 illustrates reconfigurability where point 1 shows when data reaches the traffic load percentage value.

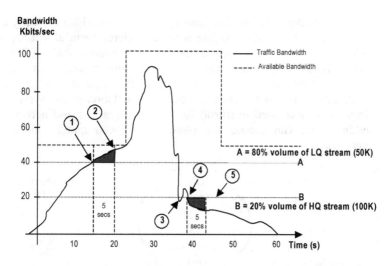

Figure 0-43: Adaptation Algorithm

The volume of data has reached the percentage value that has been set on low->high util‰. In this particular case, data volume must exceed 80% volume for a certain length of time, as in ave interval‰ before the low quality stream can move to a higher quality stream. Point 2 marks the point at which data volume has exceeded the traffic load percentage value for five seconds. The clients now receives video at the higher band rate automatically, and continues doing so open until data volume drops below a configurable level as in parameter high->low util‰. At point 3 in the diagram, traffic decreases temporarily before increasing again. Because bandwidth requirements can change suddenly like this, the algorithm waits for a period of time before readapting. In the above diagram, this value has been set to 5 seconds. At point 4, data drops below the lower traffic load percentage value (20% of 100K). Because traffic volume must remain below this threshold for a certain length of time, the client does not revert to receiving the lower quality stream until point 5 (5 seconds later) has been reached. Pseudo code representing movement between groups using the PQT technique follows:

```
IF aveThroughput_ > MaxOfCurrentGroup_ & CurrentGroup[i] <>
HQ_GROUP THEN
        CurrentGroup [i]: = CurrentGroup [i + 1]; // Move to
higher quality

IF aveThroughput_ < MinOfCurrentGroup_ & CurrentGroup[i] <>
LQ_GROUP THEN
        CurrentGroup [i]: = CurrentGroup [i - 1]; // Move to
lower quality
```

A move-up decision based on time spans can come either too early or too late. A scaling which is too late is not considered harmful if it happened with the range of a few seconds (temporarily reduced quality). A scaling which is too early can be more severe. Scaling a stream up while the congestion situation is still present causes the receiver to trigger a new scale-down and, in the extreme case, an oscillation of the system. This implies an increased overhead for both end-systems and network and, additionally, can extend the phase of reduced quality longer than necessary.

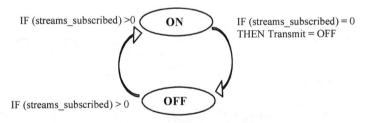

Figure 0-44: Server State Diagram

A snapshot of the state machine [Sakharov00a, Sakharov00b] for monitoring active subscribers is shown in Figure 0-44 where the server entity consists of two stages. In the ON state, the state transmits the specific media stream and if by monitoring channel subscriptions, it finds that no clients are currently subscribed to that particular stream then the source switches to the OFF state and stops transmitting that specific quality of stream. The state machine on the proxy consists of three states: In the OK state, the client receives the particular media stream in its fullest quality. If the loss rate proceeds below the loss rate threshold, there is a transition to a lower quality stream. Client's remains in the DOWN state until no more loss reports are received for a specified time span t_{up}. In this case, the protocol switches to the UP state and attempts to scale up the stream. The state machine returns to the OK state once maximum quality is reached.

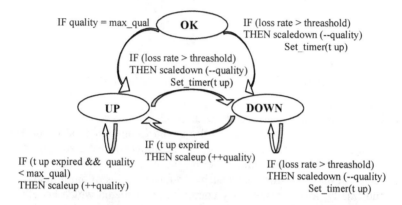

Figure 0-45: Quality Adaptation State Diagram

The primary quality transformation is conducted by PQT while finer grained quality transformations are performed by SQT using transcoding techniques. SQT compliments the PQT technique and its machinery is discussed elsewhere with regards filters etc. The algorithm for the SQT is similar to the PQT with the primary difference being SQT invokes a transcoding/filtering transformation on the existing steam rather than move to a multicast group providing higher or lower quality. The reconfiguration algorithm is illustrated (high level) in Figure 0-46. Data is lost for the duration of a protocol stack reconfiguration when transporting RTP/UDP message streams. TCP message streams do not result in lost data.

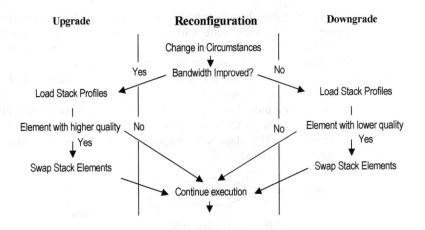

Figure 0-46: Stack Reconfiguration Algorithm

Upon activation of the reconfiguration algorithm, certain activities may take place such as transforming the media to another type in the media hierarchy; reduction of the quality of the attached media stream in order to improve down load time; altering the dimension of media which are scaled in the X, Y dimension such as audio in terms of amplitude or tone and finally, lossy or non-lossy compression of media which benefit such as text, wav, postscript etc. At present, adaptation is performed in an ad hoc manner. The limitations of such a technique are acknowledged but nonetheless useful systems can be built. It is hoped to investigate the use of a technique such as Markov Decision Processes (MDP) [Leonard96, Feinberg02] to aid adaptation decisions in the future.

4.7 Mobility Management

Chameleon utilises mobile IP, filters (or proxy) and location-based information to allow a mobile client to roam across various networks while ensuring that the mobile receives an optimum quality of service. Host mobility requires changing the stationary computers executing the service proxy. Let us assume that a student goes out of the campus with a GSM/WLAN enabled mobile device. On campus, a wireless LAN can be used but outside campus, only a cellular network can be used for connecting to the Internet. In this example, an object that reduces the frame rate of a video stream may be inserted in the protocol graph of the service proxy when the mobile moves.

Mobile IP provides the wireless network infrastructure where packets from a Correspondent Host to the Mobile Host are routed to the corresponding Home Agent (see Figure 0-47). The Home Agent looks up the address of the Mobile Host and tunnels the packet. Since all packets are sent through the Home Agent (unless route optimisation is used), the Home Agent can be a performance bottleneck when the number of Mobile Hosts increases.

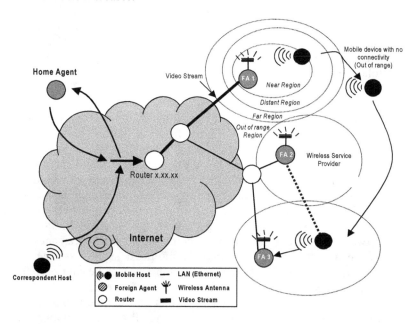

Figure 0-47: Mobility of hosts

The architecture aims to support multiple wireless technologies and thus has to be able to use multiple wireless network cells where wireless cells assume the role of the Foreign Agents. A mobile device is assigned an IP address by its current foreign agent and can be reached using that address thus the care-of address is the new address of the mobile device. The internal structure of the cells with routers and base stations is not modelled in Figure 0-47, since neither the mobile device nor the Home Agent need to know about it. The type of mobility support (e.g. Mobile IP, Cellular IP etc) used by each cell is transparent for the mobile device. In Figure 0-47, can see an example of media being streamed from the home agent to the wireless cell 1. Foreign Agent 1 (current base-station for mobile device) transmits the packets on the wireless link, while Foreign Agent 2 buffers them. When the mobile starts to moves into wireless cell 2, it registers itself and the new base-station sends a location information packet to the source base-station.

Whenever a Base Station receives a new mobile device into its network, the default is for a new local multicast group to be created for this host. A key design decision of Chameleon is that a local multicast group is always created unless the foreign network does not support multicast – then tunnel traffic through the home network. The reason for this is that local multicast traffic is more efficient than global multicast. Quite often the local multicast group is simply a mirror of a global multicast group – and is simply created so as to allow the mobile host to act as a normal host – and also for overcoming security barriers as detailed later. It is the role again of a proxy manager as to which action to take within each wireless cell.

If each service area creates a local multicast group, then a mobile host joining a group will be no different from any normal host on the same subnet therefore it may use its local Care of Address in the foreign network. The advantage here being that local multicast is more efficient. The fact that each Base Station's Proxies communicate with all the others means that the session can be continued once the mobile host moves to another service area. It can also set it up so the home network is always the default. All multicast traffic has to be tunnelled bi-directionally between the mobile host and its home agent. The advantages of this choice are that it does not require multicast support on the foreign network, and the mobile host will retain its membership as it moves around. The disadvantage is that the route is less efficient. One instance of the need for joining a multicast group through the home network is where firewalls must be traversed. Due to IP address spoofing attacks, and in accordance with the IAB [Ferg98], many routers are filtering on the source address (ingress filtering) and drop packets where source addresses do not match originating networks. Thus when a clients sends packets directly to the correspondent host in its home domain with the source IP address of the packets set to the mobile host's home IP address, the boundary router will drop these packets as they arrive from outside of the network (yet claiming to originate from within).

To overcome this problem, packets are tunnelled from the mobile host; sent through its home agent to its correspondent hosts, in a similar manner to the means to packets which are sent to the mobile host are tunnelled. This bi-directional tunnelling addresses the problem related with ingress filtering, but at increased cost. If a mobile device moves far from its home network and attempts to establish communication with a correspondent host connected to a nearby network, then packets originating from the mobile device must travel back to the home networks and then onto the correspondent host, increasing the length of this reverse path substantially.

121

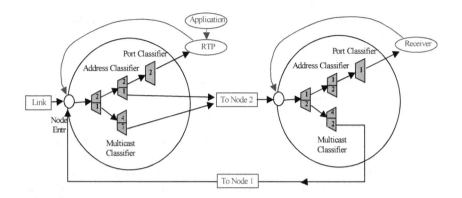

Figure 0-48: Packet flow within network layer of mobile

Figure 0-48 illustrates the flow of data through a mobile node where an address classifier is used in supporting unicast packet forwarding. A multicast classifier classifies packets according to both source and destination addresses. A Classifier object inspects packets in order to collect per-flow state. Necessary components in a streaming media application are the capture/render, encoder/decoder, framing/reassembly, and network I/O pieces. In Chameleon, these pieces are implemented in a class library adhering to the ALF principle whereby the application has to understand the interface of each piece and coordinate the buffer management and data flow between components. Each component is responsible for a particular task or function in the process of transmitting and receiving multimedia streams.

4.8   Reflective Class Loader

Inheritance alone is insufficient as far as protocol composition goes because it does not offer enough flexibility, such as the implementation of a new algorithm or the selection of among several protocol algorithms at run-time. *Reflection* also known as *open implementation* (OI) techniques [Kiczales91, Kiczales96, Kiczales97, Lopez98, Lippert99, Brichau00] allows the separation of the functional requirements from the non-functional requirements through the use of meta-object protocols. Chameleon incorporates an extended basic Reflective Java API to allow the middleware to incorporate behavioural runtime reflection. Similar work that extends Java's standard reflective API include Dalang [Welch98, Hof00], OpenJava [Chiba98, Tatsubori98, Tatsubori99,

Tatsubori00], Jasper [Nizhegorodov00, Douence00], Javaassist [Tatsubori01, Chiba98, Chiba98b Chiba00, Chiba03] and [Golm97, Golm98, Parlavantzas00 and Tramontana00] but none of these projects support mobility to the same degree.

A reconfiguration service should cause minimal execution disruption, both during normal system execution and a configuration change. Execution disruption during normal execution due to the overhead introduced by monitoring conditions can be minimised by a careful implementation while the execution disruption aspect during actual reconfiguration should affect the least amount of system activities. Chameleon uses a modified class loader to dynamically 'rewrite' a class dynamically (at run-time) or statically (at compile-time) so as to implement a simple metaobject protocol, which controls the behaviour of an underlying class. The metaobject can redefine the handling of methods in order to add desirable properties such as fault tolerance, resource reservation or security. Introspection is achieved by using the standard java.lang.reflect package that allows the insertion of new meta-classes for at run-time with no change necessary to the Java Virtual Machine. This is achieved by sub classing the existing class that the new meta-object class intends to replace, and overriding appropriate methods in the new implementation to create new meta-objects. This approach allows code in the original meta-object class to be reused by the new meta-object class, thus minimising the amount of coding required.

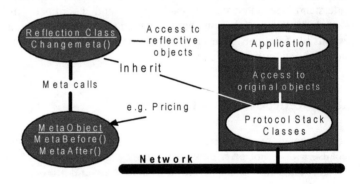

Figure 0-49: Metaobject Protocol

The Method object, which has methods to query the name, the parameter types, and the return type of the method it represents allows the inspection of the members of each protocol stack class to check if it supports a relevant protocol profile that is optimal for the current network conditions. Chameleon's reflective implementation (see figure above) thus allows the insertion of new QoS modules into the framework, which enables the java 'footprint' (i.e. number of library components) on the machine to be kept to a minimum freeing valuable memory resources. Modules may be inserted into the system, which perform data manipulation at run-time such as an image compression module for MJPEG. The framework has predefined system 'hooks' that new modules may link with, accept the data, transform it and send it along the pipeline to the next component. Chameleon intercepts method invocations of stack objects through modification of the class at the bytecode level. The java.lang.reflective class determines the interface of the original class and generates a reflective class with the same name and methods on the fly. This reflective class inherits from a metaobject class, which defines the pre & post transformations applicable to each method invocation prior to each method invocation being forwarded to the original class. The reflective class loaders to determine which Meta objects apply to which classes and methods use a Meta profile file. Metaobject classes subclass from Metaobject class and redefine the method *handlereflectivemethod()*. Each class object, which is to be dynamically reflected, must be associated with a corresponding Meta Profile file. In many instances – the same Meta Profile file is used for multiple objects. The format for the Meta Profile file is:

```
class someOriginalClass metaobjectclass-is metaClass
{
 method-detail;
 ...
 methodN-detail;
}
...
class baseClassN metaobjectclass-is metaClassN
{
 ...
}
```

Each of the method-detail lines take one of the following formats as shown in Table 0-6:

*	Allow all of the classes methods to be rewritten
method	Allow only the specified method is to be affected
* : metaParameter	The parameter metaParameter is passed to the metalevel
method : metaParameter	The parameter metaParameter is passed to specified method in metalevel

Table 0-6: Parameters used within Meta Profile File

For example a class, which allowed all methods to be rewritten, is specified in the Meta Profile file as:

```
class * metaobjectclass-is
reflect.MetaReserveBandwidth
{
  *;
}
```

After authentication, the meta-space manager calls a privileged method from the imperative interface of the meta-space it manages, to replace the current object (e.g. scheduler, NAK etc) with the new one, and initiate the object state transfer. It is then possible to dynamically change the object without interruption of service, even when there are other meta-spaces and applications running on the meta-space that is being changed. Chameleon's reflective model is coupled tightly with the dynamic stack component providing a neat principled means of creating a flexible framework for the construction of networked multimedia.

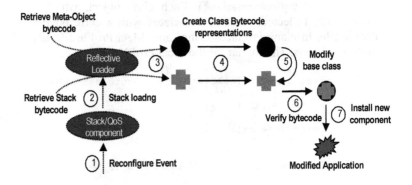

Figure 0-50: Reflective Class Loading Process

A description of the reflective class loading process is illustrated in

Figure 0-50. When a reconfiguration of a stack component event is issued by the system monitor (1), the protocol profiler will invoke the reflective class loader (2) to instantiate the loading process of the meta-object and the original base-class (3). The compile time classes are created for both classes and these objects contain the representation of the bytecode of the classes. The Reflective class loader ultimately modifies the representation of the original base class (5) and all subsequent invocations will be redirected to the meta-class. Next, a verification and consistency check will be performed by the JVM on the newly instantiated class and the new class will be loaded into the network path (7).

4.9   Protocol Profiler

The protocol profiler is responsible for connecting filters (stack layers) in the proper order. Components that form a stream are the source filter, stream sink and zero or more transform filters. The RTP source filter upon arrival of new data, detects the new source endpoint, retrieves the data and passes the data on to an RTP receive payload handler. This handler reassembles the data packets into video/audio frames in accordance with the payload specification and forwards this data to the codec, which decodes the data and passes it to the video/audio sink filter for playing. All these actions use a single thread running on a RTP source filter without necessitating any interaction on the part of the application. Figure 0-51 illustrates an instance of the protocol stack used to transport RTP data over an IP network. Once initiated, the server transmits an entire video stream as a sequence of blocks using RTP ensuring a specified gap between successive packets (rate control). The header of each packet identifies the stream name in order to uniquely identify each packet in the video stream.

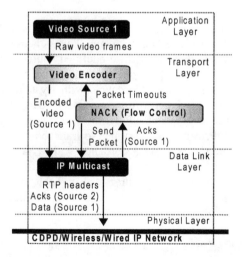

Figure 0-51: Transmitting multicast RTP data

126

The protocol profiler orchestrates all data transformations to best adapt the multicast communication for the environment at hand. A rate-limiting adapter reduces the rate at which data flows from the Server to the proxied session. Rate-limiting filters provide a mechanism for connecting low-bandwidth clients using rate-based congestion control. Traditional congestion control algorithms limit the total bandwidth of the session to that of the weakest client. Interposing a rate-limiting proxy on each of the multiple multicast groups between the low and high-bandwidth clients can alleviate this problem. This effectively splits the Chameleon session into multiple partitions. Chameleon protocol filters can be tightly coupled to the application, exploiting application-level knowledge to transform data objects. Data transformation allows the proxy to adapt the data according to the clients' device characteristics as clients may be incapable of handling certain data types such as PDAs which do not support standard formats such as AVI but instead use simple bitmap representations. Here, a filter can convert these more complex data types into representations that an unsophisticated client can understand. Media transformation also enables rate adaptation through compression, which can be lossy or non-lossy dependent on the nature of the underlying data. Video is well suited for lossy compression, since much of the resolution and colour information can be reduced or discarded, with limited degradation to the information conveyed which is particularly useful when the client devices are physically incapable of handling colour or high resolution.

4.9.1   Transcoding Media

One of the main tasks of the protocol profiler is to find the appropriate transcoder for a given media stream. This is achieved by equipping the protocol profiler database with predefined source formats and supported destination formats. The decision is made to invoke a particular transcoding format in accordance with system parameters such as available network bandwidth , processor load, screen resolution and available memory (provided by the System Monitor component, see section 4.4). Priorities are assigned to transcoding formats so that clients receive optimal quality in accordance with available system resources. These priorities are mapped in a table starting from the highest to lowest in quality (and priority) as illustrated in Table 0-7 and Table 0-8. New formats can also be added from within the transcoding class `(new Audio(AudioFormat.MPEG_RTP, 44100,16,2));`

Priority	Codec	Sampling rates	Bits/Sample	Mono/Stereo	Bit rate kbps	Computation
1	MP3	44100	16	Stereo	128	High
2	MP3	22050	16	Stereo	64	High
3	MP3	44100	16	Mono	64	High

4	DVI	22050	4	Mono	64	Low
5	u-law	8000	8	Mono	46	Low
6	DVI	11025	4	Mono	45	Low
7	MP3	22050	16	Mono	32	High
8	DVI	8000	4	Mono	32	Low
9	GSM	8000	8	Mono	13.2	Low

Table 0-7: Protocol Profiler Audio Transcoding Formats

Priority	Codec	Resolution	Frames per second	Bit rates kbps	Computation
1	MPEG	352 x 288[1]	24	280	High
2	H.263	352 x 288	24	120	Low
3	MPEG	352 x 288	12	120	High
4	H.263	352 x 288	12	50	Low
5	MPEG	176 x 144	24	120	High
6	H.263	176 x 144	24	40	Low
7	MPEG	176 x 144	12	40	High
8	H.263	176 x 144	12	50	Low
9	MPEG	88 x 72	8	30	Very Low

Table 0-8: Protocol Profiler Video Transcoding Formats

Not all formats are included here but the table should illustrate the method in place for selection of transcoding formats. For instance, using the above audio formats, a client may firstly choose to receive audio in 44 kHz, 16 bit stereo format, but the available bandwidth may diminish over time to cause the protocol profiler to transcode to 8 kHz 8 bit, u-law mono audio stream. The selection of both new audio and video transcoding formats is simplified by selecting the same priority formats from both the audio and video tables, thus a priority 6 audio format (DVI, 11 kHz, 16 bit mono) is matched with a H.263, 176x144, 12 fps video stream. The transcoding process begins when a client experiences either a reduction in available bandwidth (or over-resource utilisation) or indeed an increase in quality thus prompting an upgrading of media quality. The client is informed by the system monitor control component. The client contacts the protocol profiler to initiate the search for an appropriate new media format. The protocol profiler consults the stack repository and the audio and video transcoding format table to check on available transcoding formats as illustrated in Figure 0-52.

---

[1] The 352 X 288 (240 NTSC) format used above is used by VCD and DVD. It is also the official VHS resolution, determined by the creator of VHS, JVC.

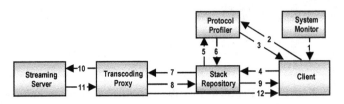

Figure 0-52: Transcoding Control Flow

Once a transcoding format is chosen, the stack repository sends a request to the transcoding proxy to forward the new transcoded stream to the client. The transcoding proxy throughout this is subscribing to the initial stream but performs a transcoding on this stream in accordance with the new mapping requested. If a media group quality shift has been requested (i.e. a PQT) then a request (as indicated by arrows 10 and 11 in above diagram) will be initiated.

4.9.2 Semantic Continuity

Reconfiguration of components in the data path provides great flexibility in responding to a range of resource variations and link failure but a fundamental problem is that such reconfiguration must preserve application semantics. The system must take into account the fact that data may have been partially processed, in transit between execution environments or lost due to node failures. Chameleon provides a solution through a combination of buffering and delayed forwarding of semantic segments at the proxy with selective retransmission of segments which are incompletely delivered (unless out of date). When a reconfiguration is triggered, the proxy begins buffering segments while continuing to transmit them. The client will monitor all segments arriving until it completely receives an output segment satisfying the property that all subsequent segments correspond only to input segments either buffered at the proxy or not yet transmitted. This is the ideal time to commence the reconfiguration and once complete the proxy will recommence transmitting starting from the first segment whose corresponding output segment was not forwarded.

4.9.3 Stack Repository

Chameleon facilitates the introduction of new stack components, which are stored on remote HTTP servers. When a stack is configured on the server, which contains a stack component not recognised by the proxy (or client), then the proxy initiates a request to retrieve this class. The web server responds to a GET command and serves a JAR file containing any classes that are needed by the proxy. To enable dynamic class loading, must set the system property 'java.rmi.server.codebase' to point to the local host name and port number of the web server. Thus when the proxy attempts to create a stack matching the server's and find that it lacks a layer, it will then check its java.rmi.server.codebase property to retrieve the bytecode from the specified codebase. Because the codebase is a URL, it will connect to the web server and retrieve the JAR file containing the new classes and load them accordingly as illustrated in Figure 0-53. These new classes are then sent through the control channel to the client that in turn reconfigures its stack.

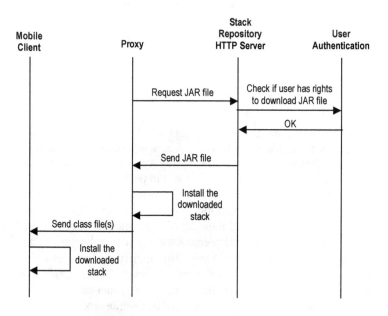

Figure 0-53: Stack download scenario

Deployment of new stack components is as simple as adding them to a jar file and placing them in a JAR file within the repository directory on a HTTP server. The Jar (Java archive) file is a collection of class files produced as a result of the Java compilation process. Each stack exists as one Jar file that is subsequently placed on the WWW server so that it can be referenced via an URL. Communication systems require matching protocol stacks to enable entities to communicate and interoperate. The downloading of required protocol stacks components from repositories encourages code sharing and reuse. The stacks are executed within

130

mobile proxies and mobile clients, playing a pivotal role in providing a mechanism that provides new services and enhances end-to-end performance and utilisation. Java virtual machine sandbox environments are deployed on proxies throughout the network to perform transcoding/filtering functions. From the mobile clients point of view, the proxy is a communications hub and message format translator, which basically awaits a set of requests to load and execute filters on behalf of clients.

In order to provide maximum reuse of protocol components, all stack components had to conform to a common interface thus allowing any stack component to be layered on top of another. There is the risk that this encourages the creation of invalid protocol stacks but it was discovered enforcing strong typing to avoid invalid stacks limited the ability of generic layer reuse. This resulted in large amounts of duplicate code.

4.9.4 Security

Java provides a number of language level security features that allow class libraries or frameworks to be secure and open at the same time. The enabling feature for this is the ability to control access to fields and methods. In Java, classes are organised in packages that possess carefully crafted external interfaces, which control access to both black box and white box classes. In addition, virtual machines divide classes into security domains based on their classloader such as the *java.net.Socket* class that uses a separate implementation object, belonging to a subclass of the *java.net.SocketImpl* class. The internal *SocketImpl* object is not visible to subclasses of the socket class therefore the *java.net.SocketImpl* class implements all functionality as protected methods, thereby allowing it to be used as a white box.

The Chameleon framework adheres to these conventions by including all classes as a single package with all black boxes being made final, which restricts access to the internal features used to implement the behaviour. White box classes are usually abstract with their behaviour being divided into user extensible features and fixed functionality. The combination of black box classes, fixed behaviour, and internal, invisible classes allows developers proper freedom when implementing protocols. New protocols can be created, but the framework conventions cannot be broken. Through careful design of explicit interfaces it possible to *extend* the framework without the possibility of breaking the conventions used by the classes provided by the framework itself. The security concerns of downloading at run-time new protocol stacks from WWW/FTP servers can be dealt with using Java's security model through the use of digitally signed bytecode and preventing remotely-loaded code from damaging the local device through the use of the sandbox model. The Chameleon system also uses a mechanism that allows the addition of security managers to the run-time system that prevent stacks from reading or modifying local files on the Host machine.

4.10   Proxies

Proxies by default are located at the home agent of each mobile client. The home agent therefore is responsible for activities such as message queuing and forwarding, access control to streaming sessions, message encryption and protocol translation among others. In order to allow the smallest possible client footprint, the home agent (and proxies) relieves the client library of most of the work of maintaining state information about active sessions, filters and general system monitoring facilities. Micro protocol stacks have been adopted to simplify the interoperability process, by allowing code for dynamic stacks to be written once and placed on repositories where they can be shared. This allows thin clients such as palm tops, WMP phones etc. to download new protocol stacks. The proxy machine executes the stacks by invoking the methods of the stack class - *create(), run(), reconfigure(),* and *terminate().* When a proxy receives a create call, it initialises the stack to be loaded by calling the *create()* method of the stack class. The class files of the Jar file contain the machine independent byte codes for the requested stack, which are loaded by the Proxy Server. The proxy instantiates an instance of the stack object, and calls the *create()* method of the stack class to initialise the downloaded stack. The *create()* request is used to create the new QOS stack. A number of acceptance checks are performed on the new stack including validation, loading, permission and security checks. If the stack is acceptable, it is then loaded and awaits a *run()* call. As stacks can be loaded by URL reference, the proxy may use a local http cache for stack libraries that have already been loaded. The *run()* method starts execution of the stack and this is where most of the code for networking and filtering is encapsulated. Arguments are commonly appended to this request that can be used to specify hostnames, or transcoders to be used or further downloaded. When a proxy no longer needs a stack, it invokes a *terminate()* method. A stack may need to be terminated if it refuses to terminate naturally.

Proxies/Filters are central to the workings of Chameleon, and it is envisaged that in a typical end-to-end communication path, one or more proxies will be used as service points to enhance performance and provide services tailored to applications. The application logic of the client is split between the mobile client and the proxy to adapt to the dynamic wireless environment and to address the limitations of the portable device.

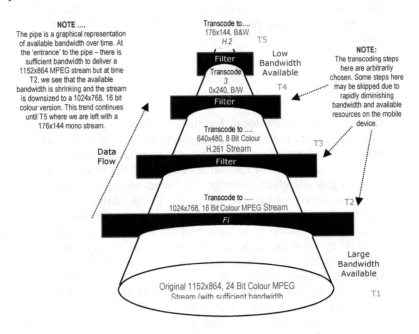

Figure 0-54 : The role of filters in transcoding media

Chameleon provides the components of a flexible and general-purpose runtime infrastructure to facilitate the rapid development and deployment of adaptive mobile applications (see Figure 0-54). The architecture enables a piece of applications on a mobile device to be executed efficiently even when the device is resource poor. Clients send multicast messages to a well-known multicast address in order to locate proxies as illustrated in Figure 0-55. Proxies and clients communicate with TCP, RTP or UDP. TCP is used to send control requests. Media can be transported by reliable and unreliable stream protocols with and without RTP. The reliable protocol is used for sending texts and control information, and the unreliable protocol is used for video and audio.

133

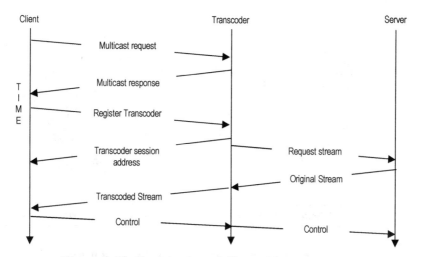

Figure 0-55: Registration of client with transcoder

A key issue in the engineering of any mobile stack model is the topological placement of the filtering components. A filtering operation can be performed either within the network infrastructure or at the network edge (e.g. end-systems and gateways). A fundamental feature of Chameleon is that it may be deployed on the 'edge' of the existing Internet. Internet Service Providers (ISPs) will support servers at their sites to run code of a third parties choice. Such boxes are being placed inside ISP's networks, to help with content delivery such as Real System Server Professional add-on products[1]. Large central servers according to queuing theory are more efficient in terms of cost and utilisation than collections of smaller servers thus pointing to another motivation to co-locating proxies with ISPs in order to achieve valuable economies of scale. This leads to an interesting observation in that perhaps others have awakened to the benefits of placing generic boxes, which enable the dynamic execution of application level services rather than "task specific boxes",

---

[1] http://www.realnetworks.com/products/servers/plus/specifications.html

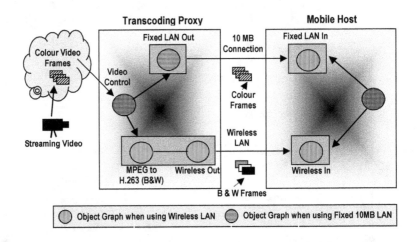

Figure 0-56: Transcoding video via a proxy

Proxies perform their work through the use of stream filters. A primary function of each proxy is to regulate the output bandwidth used by the transcoded packet streams by generating variable frame rate output (e.g. dropping frames) to meet a given rate constraint. Stacks encapsulate end-to-end delivery therefore they perform the compression, transmission and decompression of media. A host may be connected to a high performance fixed network receiving broadcast quality MPEG video with little or no perceived degradation in quality. At some point it may move to a wireless connection with reduced bandwidth and increased jitter in traffic as illustrated in Figure 0-56. In this scenario, a transcoder proxy receives the video stream on a LAN backbone at 10MB, performs a transformation on the stream before forwarding to a client connected to a 2MB wireless network. To address this change in connectivity invoke the meta-interface of the binding object to locate the video compression object and cause it to reduce the bandwidth of compressed video. At some later stage the host may move out of range of the WLAN base station and accordingly a GSM dialup link is invoked.

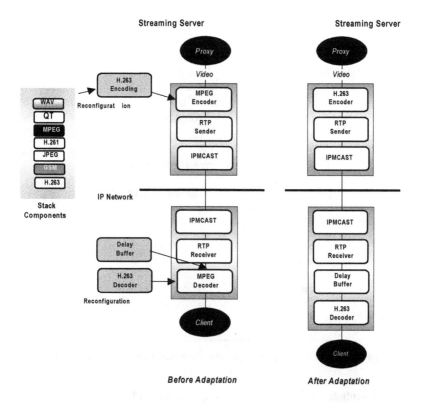

Figure 0-57: Run-time Adaptation of Encoders/decoders

It may now be more sensible to replace the video compressor object with one more suited to low-bandwidth such as an H.263 encoder object rather than simply altering the MPEG compression parameters (See Figure 0-57). A Delay Buffer component inserted directly on top of the bottommost layer (e.g. UDP) can be used to delay messages. Each stack component will accept an input stream (where the original data came from), an output stream (where the transcoded data is to be sent) and a list of track formats of which the input data will be transcoded to. The actual algorithm to transcode data is as follows:

1. Create and configure a transcoder for each input stream.
2. Parse the output media locater and determine the content descriptor based on the media.
3. Set the output content descriptor on the transcoder
4. Match each track format against the supported formats and commence transcoding.
5. Obtain the output *DataSource* and use the output media locator to start a *DataSink*.

136

6. Wait for the DataSink's *EndOfStreamEvent* and close the transcoder and *DataSink*.

The basic process generally involves transcoding from some set of supported input formats to a different set of output formats. Figure 0-58 illustrates the range of parameters available to the transcode object for transcoding a video stream into a different (lower-rate) bit.

Transcode -o <output > -a <audio format> -v <video format> <input > where
<audio format> is of the form:
[encoding]/[rate]/[sizeInBits]/[channels]/[big|little]/[signed|unsigned]
and <video format> is of the form:
[encoding]/[rate]/[sizeInBits]/[channels]/[big|little]/[signed|unsigned]

encoding - A String that describes the encoding type for this VideoFormat
size - The size of a video frame.
maxDataLength - The maximum length of a data chunk.
dataType - The type of data e.g. byte array.
frameRate - The frame rate.

Figure 0-58: Transcoding parameters for MPEG stack component

The eventual goal is to support all possible combinations of efficient transcoding operations, the current implementation supports transcoding only from the MPEG and H.263 input formats to the H.261, MPEG and H.263 output formats. Streams that are not from this group (or those on non-rate controlled sessions) are simply tunnelled through the transcoder.

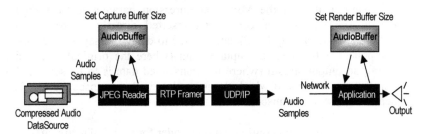

Figure 0-59: Reconfiguring Audio Buffers

In addition to audio and video transcoding, Chameleon implements data manipulation in the form of buffer control. Smaller buffer sizes will in some situations reduce latency but will have a tendency for more severe audio break-up. The variable factors in this situation are dependent on processing power, network traffic among others. An AudioBufferControl component sets the capture buffer size while media players set the render buffer size (see Figure 0-59). Parameters to the AudioBufferControl class are as follows: *AudioBufferControl [-c <capture buf size in millisecs>] [-r <render buf size in millisecs>] <Input Stream>*.

An early version of Chameleon took the view that it was better to build large user data packets in the application layer and let IP perform the fragmentation and defragmentation automatically, but audio-video was frequently delayed on low bandwidth links when large PDU sizes were used (>576 bytes) therefore to avoid fragmentation, adopted an MTU size of 576 bytes. This resulted in maximum throughput for continuous media over dynamically fluctuating heterogeneous link bandwidths. Packet lengths significantly affect the delay incurred in filling up the packet with voice/sound samples. The larger the packet, the longer the delay to accumulate sufficient voice samples, which can be generated at a very low rate. The delay when dealing with a 6.2 kb G.723.1 codec and 1500 byte sized packets was a 3.75 second round-trip. This is not acceptable where the end-to-end round trip for telephone conversations should be limited to 100-200ms [Kitavaks91]. The packet size was also limited in order to prevent larger packets introducing significant delay in other streams especially on multiplexed links. Another reason might be for the case of a voice-annotated slide show where the voice packets must be limited in size so as to intersperse with slide-show packets. Our MTU policy was influenced in part by the phenomenon discovered by [Talley97] that decreasing the message rate sometimes increases the number of packets on a hop. This can occur when the packaging reaches the point where the nth frame is successfully packaged into a packet, but frame $n + 1$ frame causes the message size to exceed the packet's capacity and require an additional packet. This situation occurs whenever $n + 1$ frames are packaged in a single message. Even though the generated packets on a hop may not always decrease with decreases in the message rate, as message size increases, the packaging of frames into packets becomes increasingly efficient and the packet rate generally decreases.

4.11   Client Applications

Clients subscribe to multicast groups to receive media packets. Chameleon extracts the different media types from within the application and creates suitable run-time protocol stacks to enable streamlined transport communication to be invoked. The Chameleon middleware is responsible for re-assembling the flows at the receiver. Code to set up a client is shown in Figure 0-60.

```
// create a simple client for the corresponding unreliable
stream
final Stack stack = new Stack("FRAG:RTP:MCAST");
final         ChameleonURL         =         url         =
URLFactory.create("MyBroadcast",   "1.1",   "/connections",
www.infm.ulst.ac.uk/~kevin/chameleon, 7634);

Try {
        Stack.registerclient (url);
} catch {Exception e) {
        log.error ("register client failed:" + e);
```

Figure 0-60: Sample code to create a client

A sample client media player is illustrated in Figure 0-60. This is a Java application, which is bundled with the chameleon middleware that enables one to view streams from the Multi-stream server. Figure 0-62 shows the proxy viewer, which the administrator can use to 'snoop' on streams passing through on their way to mobile clients. This is similar to the client player and will only differ in actual quality of video displayed (as the stream a proxy views may be of a better quality than that which is transcoded and passed to the mobile client).

Figure 0-61: Client Player   **Figure 0-62: Proxy Player**

Chameleon also has a GUI for snooping on streams as they pass through a proxy. Default profiles can be over-ridden using the GUI and new formats specified. For example, in Figure 0-63, the outgoing Audio HQ stream (top left of diagram) could be changed from DVI at 46Kb/Sec to PCM encoding at 78 Kb/Sec. This is an example of user level QoS being applied.

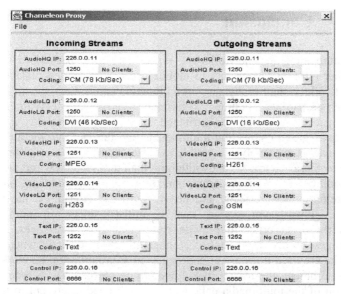

Figure 0-63: Proxy GUI for Stream Transformations

With application controlled QoS (as in Chameleon), different applications running on the same system may have very different adaptive behaviour when QoS variations occur. Some of these may consume a considerable amount of system resources to perform their desired adaptive behaviour, (provided resources are available) while others may not perform any adaptation at all. The GUI above provides system administrators with a snapshot of the overall server-proxy-client stream transformations at any time during an active session.

5   Conclusion

Future integrated networks are expected to offer packetised voice and other multimedia services to mobile users over wireless links. Wireless networks cannot easily support multimedia applications due to the media high probability, burstiness, persistence of errors and delay jitter suffered by packets. This can be attributed to traditional variability in queuing delays due to connection multiplexing and the fact that the air interface channel is extremely susceptible to noise, fading and interference effects. Thus there is a lower bandwidth limit than that seen on wired links. Node mobility imposes extra requirements on routing introducing handovers, motion-induced effects, and heavy routing overhead. At the transport layer, lower level services must be utilised, under minimum protocol overhead, in order to provide end-to-end QoS guarantees. Applying congestion and flow control is inherently more difficult when wireless links are used not only at end nodes, as in a multi-hop network. In this environment, the transport will need to be presented with a compressed form of the multimedia data, and thus will need to present to the upper layers a reliable service. Furthermore, seamless interaction with wired peers is generally required. Widely accepted transport protocols have been written for wired networks and will need middleware to achieve acceptable performance. This book has documented one solution to the above through a middleware equipped with adaptation mechanisms, which strive to achieve maximum throughput to resource-constrained devices in heterogeneous environments. The framework approaches the problem of streaming different qualities of media to devices over a range of networking infrastructures in two broad directions:

- To provide a reflective adaptable object oriented framework API for building multimedia applications through the use of micro-protocols where active proxy/filters can be deployed throughout the network to help overcome congestion, offload processing power, and provide transcoding services (SQT) for low powered devices.

- To provide an algorithm (PQT) utilising multiple multicast media groups, which allow heterogeneous clients to select and combine optimal prioritised streams, to cope with fluctuating network conditions in a coarser manner than transcoding services mentioned earlier.

Trends in system design, which support the need for Chameleon, include the increasing popularity of component architectures that reduce development time and offer freedom with choice of components. This allows alternative functionality to be deployed in various scenarios to combat differing QoS needs. Another trend is introspection, which provides run-time system information allowing applications to examine their environment and act accordingly. The middleware provides an infrastructure for building adaptive applications that can deal with drastic environment changes. The Internet with its multiple standards and interconnection of components such as decoders, middleware, databases etc. deserves more than a plug, try and play mentality. The introduction of mobility increases the complexity due to the proliferation in possible actions. A key goal was to provide a principled means of allowing a system to be adapted throughout its lifetime with the minimum of effort thus the principles of reflection provides a means of coping with change in a computing system as it allows access to the implementation in a principled manner. The concept of filtering media streams becomes interesting if filter operations can be performed without significant overhead. The results of the experiments conducted here show that filter algorithms can be applied in real-time if proper application layering is used. Transcoding of media can be achieved on current off-the-shelf hardware only with a limited amount of subscribing clients. Multi-processor machine with RAID or other special hardware for transcoding would lead to increased support for multiple clients. One drawback of transcoding is that the filtering is not suitable to address scenarios which require loss-less transmission of steams. Filtering, by design, is a lossy mechanism which reduces date rate and ultimately quality. Chameleon can coordinate packet filtering with error detection and FEC techniques to cope with packet losses due to pure transmission errors.

Chameleon's protocol stacks enable a developer to select a particular protocol profile at bind-time where each protocol profile is built from a rich library of protocol modules including UDP, packet loss detection, data encryption, TCP, Multicast among others. Communicating objects are represented as object graphs that together realise the required behaviour built upon the IP service offered by the host computer. All protocols down to device driver level can be implemented at the user level, providing the maximum potential for configurability. The reflective API allows the dynamic reorganisation of stacks whilst 'data is flowing'. The use of the object oriented language Java allowed the use of object-level design patterns, which resulted in the framework becoming generic and extensible (the reflective approach here is similar to the Decorator Pattern [Gamma95]). Another characteristic of Chameleon is that the source and the set of receivers are very loosely coupled in their control and data exchange interactions interacting indirectly through an independent entity, multicast groups. This open-loop property is important for any multicast communication protocol that is expected to scale to large numbers of receivers. Chameleon differs from the layered encoding schemes such as Receiver-Driven Layered Multicast because it explicitly addresses link sharing and de-couples the bandwidth of the unit of network control (the multicast group) from any video encoding scheme. Java enabled mobile devices, which are able to execute Java application software faster and more responsively are on the increase due largely due to Java byte code instruction interpretation being done on the fly in hardware using advanced optimisation techniques. Executing Chameleon software in hardware enhanced systems can lead to ten-fold CPU utilisation efficiency, as the CPU is not burdened with interpreting or compiling Java byte code instructions. Ultimately in the near future performance gains and significant reduction in overall system power consumption can be expected as a result of increased computing power in the mass market.

The framework presented here is an extension of the common client-server architecture where one host (server) provides certain services or application of which another host or device (client) makes use. The proxy (gateway or mediator) is another host logically placed between the server and client in order to mediate, route, or otherwise facilitate communication between them. The client-proxy-server approach can be implemented at different levels in the protocol stack. When it is applied to the network or transport layers, the result is a mobility-transparent solution. Mobility solutions that are transparent to the application are desirable because of the reduced impact to the applications in that they allow most of the fixed network (Internet) to remain architecturally unchanged. They also allow traditional protocols and software application designs to operate as they always have. They support the use of new and optimised protocols between the fixed network and mobile clients. Given the fact that fixed servers have thus far been several orders of magnitude more powerful than mobile devices, these solutions also allow the use of such powerful servers to perform resource intensive filtering, compression and protocol conversion tasks. This framework generalises the data stream adaptation approach where multiple adaptive proxies filter the stream along the path. Each adaptive proxy further decomposed into an Event Manager (EM), a System Monitor (SM), and many more components which attempt to transform the data stream and make requests for additional resources as required. Proxies are placed at the edge between the traditional fixed network and the mobile network in order to modify and adapt content to suit each mobile device. The greater computational and system resources of the gateway machine is used to "digest" the data stream and pass on a version which is suited to the wireless bandwidth and end-user device capabilities. While transparent support for wireless networks and mobile clients is favoured as a non-intrusive method for integrating mobile networks with fixed networks, it can also be argued that the mobility cannot be fully exploited without the application's involvement. Therefore, the load created for proxies carrying out the transparent adaptation may create a scalability problem as wireless devices increase in number. The proxy model can result in large amounts of data being passed over networks, followed by resource intensive computations on the proxies, which can ultimately result in data being discarded or modified. This suggests that end-to-end solutions that involve the application in the adaptation decisions may provide better scalability. Infrastructure solutions that are meant to be transparent can also be designed to address scalability. This leads to the concept of applications that are aware of the mobile environment they are operating in. The solution framework presented here achieves this by structuring mobility support into a middleware layer that the application can use to be notified of, and respond to, network changes. The use of a middleware approach will provide greater flexibility for application development alongside the useful functions that proxies provide as previously mentioned. Applications can be developed for many purposes and as long as they call on the functions of the middleware they can support mobility.

Please note that the Chameleon framework discussed here is available for free to third-parties for non-commercial use. Please email the author at kj.curran@ulster.ac.uk for further details.

References

[2K03] 2K: An Operation System for the Next Millennium, http://choices.cs.uiuc.edu/2k.

[3GPP-03] 3GPP, Open services architecture. Application Programming Interface, 3G TR 29.998, http://www.3gpp.org, 2003

[Abbott93] M. Abbott and L. Peterson, Increasing Network Throughput by Integrating Protocol Layers, ACM Transactions on Networking, vol. 1, October 1993.

[Almeroth00] K. Almeroth, The evolution of multicast: From the MBone to inter-domain multicast to Internet2 deployment. IEEE Network, January/February 2000.

[Amir95] E. Amir, S. McCanne, and H. Zhang. An application level video gateway. In ACM Multimedia 95, November 1995

[Amir97] Amir E.,RTPGW: An Application Level RTP Gateway, IEEE Network Journal, July 1997.

[Amir01] F Amir and S McCanne and M Verter1i. A layered DCT coder for Internet Video. Proceedings of the IEEE Conference on Image Processing ICIP'96, September 2001

[Anagnostov91] Anagnostov, M.E., Cuthbert, L., Lyratzis, T.D., Pitts, J.M.: Economic Evaluation of a Mature ATM Network. Journal of Selected Areas in Communications 10, pp. 1503-1509, 1991

[Anastasiadis02] Stergios V. Anastasiadis and Kenneth C. Sevcik and Michael Stumm. Maximizing Throughput in Replicated Disk Striping of Variable Bit-Rate Streams, http://citeseer.nj.nec.com/anastasiadis02maximizing.html

[Apteker95] Apteker, R. Video acceptability and frame rate, Witwatersrand University, IEEE Multimedia, 1995

[Arcangeli00] Jean-Paul Arcangeli, Laetitia Bray, Annie Marcoux, Christine Maurel, and Frédéric Migeon. Reflective Actors for Mobile Agents Programming. ECOOP'2000 Workshop on Reflection and Metalevel Architectures, Sophia Antipolis and Cannes, France June 12 - 16, 2000

[Aurrecoechea98] Aurrecoechea98, C., Campbell, A. and Hauw, L. A survey of QoS architectures, ACM/Springer Verlag Multimedia Systems Journal, Special Issue on QoS architecture, Vol. 6, No. 3, pp. 138-151, May 1998

[Austerberry02] David Austerberry. The Technology of Video and Audio Streaming. Focal Press; May, 2002

[Bahl97] Bahl, P., Chlamtec, I. and Farago, A. Optimizing Resource Utilisation in Wireless Mutimedia Networks. Proceedings of IEEE International Conference on Communications, Quebec, Canada, 1997

[Bajaj98] S. Bajaj, L. Breslau, and S. Shenker. Uniform versus priority dropping for layered video. Proc. of ACM Sigcomm, (Vancouver, British Columbia, Canada), pp. 131–143, September 1998.

[Baker98] Baker, F., Brim, S., Li, T., Kastenholz, F., Jagannath, J. and Renwick, J. IP Precedence in Differentiated Services Using the Assured Service, Internet Draft draft-diff-serv-precedence-00.txt, April 1998.

[Balasub03] Balasubramanian, Vidhya, Venkatasubramanian, Nalini. Server transcoding of multimedia information for cross disability access. ACM/SPIE Conf on Multimedia Computing and Networking,, 2003.

[Badrinath00] B. R. Badrinath, A. Fox, L. Kleinrock, G. Popek, and M. Satyanarayanan, A Conceptual Framework for Network and Client Adaptation. Mobile Networks & Applications. Vol. 5, No. 4, 2000

[Barbier98] Barbier, F., Briand, H., Dano, B., Rideau, S. The executability of Object Oriented Finite State Machines, Journal of object oriented programming, July/August 1998

[Barbounakis00] Barbounakis, I, Stavroulakis, P., and Gardiner, J., G. General Aspects of Digital Technologies for Wireless Local Loops. International Journal of Communication Systems, May, 13(3), 187-206, 2000

[Barnes03] J. Barnes, C. Rizos, J. Wang, D. Small, G. Voigt, N. Gambale. LocataNet: A New Positioning Technology for High Precision Indoor and Outdoor Positioning. Proc. of (Institute of Navigation) ION GPS/GNSS 2003 - Oregon Convention Center, Portland, Oregon, September 9-12, 2003

[Bates02] Juliet Bates. Optimizing Voice Transmission in ATM/IP Mobile Networks (McGraw-Hill Telecom Engineering), ISBN: 0071395946, May, 2002

[Becker00] Becker, C., Geihs, K. Generic QoS support for CORBA. In Proceedings of the 5th IEEE Symposium on Computers and Communications (ISCC'2000) Antibes, France 2000

[Becker03] Becker C., Schiele, G., Gubbels, H., Rothermel, K. BASE - A Micro-broker-based Middleware For Pervasive Computing. Proceedings of the IEEE International Conference on Pervasive Computing and Communication (PerCom), Fort Worth, USA, July 2003

[Bellavista01] Bellavista, P., Corradi, A. and Stefanelli, C. Mobile Agent Middleware for Mobile Computing. IEEE Computer, Vol. 34, No. 3, Mar. 2001.

[Bellavista03] Bellavista, P., Corradi, A., and Stefanelli, C. Mobile Agent Middleware for Mobile Computing. Computer, vol. 34, no. 3, 2001, pp. 73--81.

146

[Berkeley02] Berkeley Plateau Multimedia Research Group (BMRC), Berkeley University, http://bmrc.berkeley.edu/frame/research/mpeg/

[Bershad90] B. Bershad, T. Anderson, E. Lazowska, and H. Levy. Lightweight Remote Procedure Call. ACM Transactions on Computer Systems (TOCS), 8(1), pp. 37-55, 1990.

[Bettati95] R. Bettati, D. Ferrari, A. Gupta, W. Heffner, W. Howe, M. Moran, Q. Nguyen, R. Yavatkar. Connection establishment for multi-party real-time communication. Proceedings of the Fifth International Workshop on Network and Operating System Support for Digital Audio and Video , Durham, NH, April 1995.

[Betz94] Betz, M, Interoperable Objects, Dr. Dobb's Journal, pp.45-52, November 1994.

[Bianchi00] Bianchi, G., Campbell, H. Programmable MAC framework for utility based Quality of Service support. IEEE newspaper on selected areas in communications, February 2000.

[Bingham00] Bingham, J. ADSL, VDSL, and Multicarrier Modulation. Wiley-Interscience; 1st edition ISBN: 0471290998, January 2000

[Blair97] Gordon Blair, Geo Coulson and Tom Fitzpatrick. A software architecture for distributed adaptive multimedia systems. In Proceedings of NOSSDAV, Missouri, USA, pp.259-273, May 19-21, 1997.

[Blair98] Gordon Blair, Geo Coulson, Philippe Robin, and Michael Papathomas. An Architecture for Next Generation Middleware . In Proceedings of Middleware '98, Lake District, England, November 1998.

[Blair00] Blair, G., Coulson, G., Robin, P. and Papathomas, M. An architecture for next generation middleware. Proceedings of Middleware 2000, Lake District, UK, 2000

[Blair00b] Blair, G., Andersen, A., Blair, L., Coulson, G. and Gancedo, D.S. Supporting dynamic QoS management functions in a reflective middleware. IEE Proceedings Software 2000

[Blair01] Blair, G.S., Coulson, G., Andersen, A et al. The design and implementation of OpenORB version 2, IEEE Distributed Systems Online Journal 2 (6) 2001

[Blair02] Gordon S. Blair, Geoff Coulson, Michael Clarke and Nikos Parlavantzas, Performance and Integrity in the {OpenORB} Reflective middleware, Lecture Notes in Computer Science, vol 2192, pp. 268-276, 2001

[Bolot94] Bolot, J.-C., Turletti, T., and Wakeman, I., Scalable Feedback Control for Multicast Video Distribution in the Internet, ACM SIGCOMM '94, Sept.

1994.

[Bolton01] Bolton, Fintan. Pure CORBA. Sams Publishing - ISBN: 0672318121; 1st edition, July 2001

[Bosch95] Bosch, J. Abstracting Object State, Pattern languages of programming design 3, edited by Martin/Riehle/Buschmann, Addison Wesley 1998

[Bouillet02] E. Bouillet and D. Mitra and K. Ramakrishnan. The Structure and Management of Service Level Agreement in Networks. IEEE Journal on Selected Areas in Communications, Vol. 20, No. 4, May 2002

[Braden02] R. Braden and T. Faber and M. Handley. From Protocol Stack to Protocol Heap - Role-Based Architecture. First Workshop on Hot Topics in Networks (HotNets-I), 2002

[Bradshaw01] M. Bradshaw, B. Wang, S. Sen, L. Gao, J. Kurose, P. Shenoy and D. Towsley. Periodic Broadcast and Patching Services - Implementation, Measurements, and Analysis in an Internet Streaming Video Testbed, Proc of ACM Multimedia Systems 2001, October 2001

[Braun99] Braun, T., Castellucia, C., and Stattenberger, G. An analysis of the Diffserv approach in Mobile environments. IWQiM 1999.

[Broersma03] Broersma, Matthew. Microsoft screens Smart Displays. CNETAsia, http://asia.cnet.com/newstech/systems/0,39001153,39100085,00.htm, December 2 2002

[Brichau00] Johan Brichau. Declarative Meta-Programming for a Language Extensibility Mechanism. ECOOP'2000 Workshop on Reflection and Metalevel Architectures, Sophia Antipolis and Cannes, France June 12 - 16, 2000

[Bruneton00] Eric Bruneton, and Michel Riveill. Reflective Implementation of non-functional Properties with the JavaPod Component Platform. ECOOP'2000 Workshop on Reflection and Metalevel Architectures, Sophia Antipolis and Cannes, France June 12 - 16, 2000

[Burrows94] Burrows, M and Wheeler, D. A block-sorting lossless data compression algorithm, Technical Report SRC-124, Digital Systems Research Center, May 1994

[Buschmann96] Frank Buschmann, Regine Meunier, Hans Rohnert, Peter Sommerlad, Michael Stal. Pattern-Oriented Software Architecture - A System of Patterns. Wiley 1996. ISBN 0-471-95869-7.

[Campadello00] S. Campadello et al., Using Mobile and Intelligent Agents to Support Nomadic Users. Proc. ICIN'2000, January 2000

[Campbell98] Campbell, A., Coulson, G, Gracia, F., Integrated Quality Of

Service For Multimedia Communications, Proc. IEEE INFOCOMM'93, pg 732-39, Mar 1993

[Campbell95] Campbell, R.H., S-M. Tan, "μChoices: An Object Oriented Multimedia Operating System". Proceedings of the Fifth Workshop on Hot Topics in Operating Systems, Orcas Island WA, USA, May 1995.

[Campbell97] Campbell, A.T., Coulson G., and D. Hutchison. Supporting Adaptive Flows in Quality of Service Architecture, Multimedia Systems Journal, Special Issue on QoS Architecture, 1997

[Campbell97b] Campbell, A. Mobiware: QOS-aware middleware for mobile multimedia communications. Seventh International Conference on High Performance Networks (HPN'97), April-2 May 1997

[Carlson02] Carlson, R. ANL Web100 based Network Configuration Tester http://miranda.ctd.anl.gov:7123/. The Java applet is available freely at ftp://achilles.ctd.anl.gov/pub/web100.

[Cassidy92] Cassidy, D., Heisenberg. Uncertainty and the Quantum Revolution. *Scientific American, 266* (May 1992), 106-112.

[Cetin01] C. Cetinkaya, V. Kanodia, and E. Knightly. "Scalable Services via Egress Admission Control," in IEEE Transactions on Multimedia: Special Issue on Multimedia over IP, 3(1):69-81, March 2001.

[Chakravorty02] R. Chakravorty, I. Pratt. WWW Performance over GPRS. Fourth IEEE Conference on Mobile and Wireless Communications Networks (MWCN 2002), Stockholm, Sweden, September 9-11, 2002

[Chapweske00] Chapweske, Justin. Forward Error Correction Performance. Downloadable report located in FEC Open Source Zip Archive at http://onionnetworks.com/downloads/fec-1.0.3.zip   (fec-1.0.3/docs/fecperformance.ps)

[Chandra01] Surendar Chandra, Ashish Gehani, Carla Schlatter Ellis, and Amin Vahdat. Transcoding characteristics of web images. Proceedings of SPIE Multimedia Computing and Networking Conf, Jan 2001.

[Charzinski00] Charzinski, J.: "Internet Traffic Measurement and Modelling", IEEE ATM Workshop 2000 and 3rd Int. Conf. on ATM (ICATM'2000), Heidelberg, Germany, 26-29 June, 2000

[Chen01] Chen, Y., Huale, H., Jana, R., John, S., Wei. B. Personalised Multimedia Services using a Mobile Service Platform. Proc. Of the first ACM Workshop on Wireless Mobile Internet (WMI 2001), July 2001

[Cheung94] S. Y. Cheung, M. H. Ammar. Using Destination Set Grouping to Improve the Performance of Window-controlled Multipoint Connections. GIT, College of Computing, Report GIT-CC-94-32, Aug 1994.

[Cheung95] S. Y. Cheung, M. H. Ammar and X. Li. On the use of Destination Set Grouping to Improve fairness in Multicast Video distribution. Georgia Institute of Technology report GIT-CC-95/25. July 18, 1995.

[Cheung96] S.Y. Cheung, M.H. Ammar, and X. Li. On the use of destination set grouping to improve fairness in multicast video distribution", Tech. Rep. GIT-CC-95-25, Georgia Insitute of Technology, Atlanta, GA, Proceedings of the IEEE INFOCOM '96, July 1995

[Chiba98] Chiba, Shigeru and Tatsubori, Michiaki. Yet Another java.lang.Class. ECOOP'98 Workshop on Reflective Object-Oriented Programming and Systems, Brussels, Belgium, July 20, 1998.

[Chiba98b] Shigeru Chiba. Javassist --- A Reflection-based Programming Wizard for Java. In Proceedings of the ACM OOPSLA'98 Workshop on Reflective Programming in C++ and Java October, 1998.

[Chiba00] Shigeru Chiba. Load-time Structural Reflection in Java. ECOOP 2000 -- Object-Oriented Programming, LNCS 1850, Springer Verlag, page 313-336, 2000.

[Chiba03] Shigeru Chiba, Yoshiki Sato, and Michiaki Tatsubori. Using HotSwap for Implementing Dynamic AOP Systems. ECOOP'03 Workshop on Advancing the State of the Art in Runtime Inspection (ASARTI), July 21st, 2003.

[Cheriton95] Cheriton, David. Dissemination-Oriented Communication Systems, Stanford University, Tech Report, 1995

[Cheshire00] Cheshire, S. For every network service there's an equal and opposite network disservice. http://rescomp.stanford.edu/~cheshire/rants/networkdynamics.html

[Cisco00] Label and Tag Switching. Document available from Cisco Systems at http://www.cisco.com/warp/public/cc/techno/protocol/tgth/index.shtml

[Clark90] Clark, D.D, Tennenhouse, D.L. Architectural considerations for a new generation of protocols. Proc. ACM SIGCOMM'90, pp. 200-208, September 1990

[Clark96] David Clark and Joseph Pasquale. Strategic Directions in Networks and Telecommunications. ACM Computing Survey. 28, No. 4, 679-690 (December 1996)

[Clark97]. Clark, D., Lambert, M., Zhang, L., NETBLT: A bulk data transfer protocol, RFC 998, Mar 1997

[Clarke98] Clarke, M. and Coulson, G., DEIMOS: An Architecture for Dynamically Extensible Operating Systems, Proc. 4th ICCDS98, Annapolis, Maryland USA, May 1998 pp145-155.

[Cobley98] P Francis-Cobley and N Davies. Performance Implications of QoS Mapping in Heterogeneous Networks Involving ATM. Proc. ICATM '98, IEEE International Conference on ATM, pp529-535, June 1998.

[Coad95] Peter Coad, David North & Mark Mayfield. Object Models: Strategies, Patterns, & Applications, Prentice Hall. 1995.

[Coady03] Yvonne Coady and Gregor Kiczales. Back to the Future: A Retroactive Study of Aspect Evolution in Operating System Code In Proceedings of Aspect Oriented Systems Development AOSD'03, 2003.

[Corman01] Corman, David, WSOA–Weapon Systems Open Architecture Demonstration–Using Emerging Open System Architecture Standards to Enable Innovative Techniques for Time Critical Target (TCT) Prosecution, 20th Digital Avionics Systems Conference (DASC), Daytona Beach, Florida, IEEE/AIAA, October 2001

[Costa00] Fábio Costa, and Gordon S. Blair. The Role of Meta-Information Management in Reflective Middleware. ECOOP'2000 Workshop on Reflection and Metalevel Architectures, Sophia Antipolis and Cannes, France June 12 - 16, 2000

[Couloris01] G. Coulouris, J. Dollimore and T. Kindberg. Distributed Systems Concepts and Design. Pearson Education Limited, Essex, UK, 2001.

[Coulson99] Coulson, et al. Supporting mobile multimedia applications through adaptive middleware. IEEE j. selected areas in communications Vol 7 No 9 Sept 1999

[Crawley98] Crawley, E., Berger, L., Berson, S., Baker, F., Borden and M., Krawczyk, J. A Framework for Integrated Services and RSVP over ATM, RFC 2382, August 1998

[Cuthbert93] Cuthbert, L.G. and Sapanel, J.C. ATM: The Broadband Telecommunications Solution (IEE Telecommunications Series, No 29) Inspec/IEE, ISBN: 0852968159, June 1993

[Danthine93] Danthine, A., Bonaventure, O., From Best-Effort to Enhanced QoS, RACE 2060, CEC Deliverable No R2060/Ulg/CIO/DS/P/004/bl, Jul 1993

[Davie00] Bruce Davie, Yakov Rekhter. MPLS - Technology and Applications, Morgan Kaufmann, 2000

[Deering89] Deering, S., Host Extensions for IP Multicasting. STD 5, RFC 1112, August 1989.

[Deering90] Deering, S., Cheriton, D. Multicast routing in Internetworks and Extended LANs. ACM Transactions on Computer Systems, Vol. 8, No. 2, pp. 85-110, May 1990

[Deering99] Deering, S. IP – the next generation. Computer Magazine, (Vol. 32, No. 4), pp11-17, April 1999

[Defago03] X. Défago, N. Hayashibara, and T. Katayama.. On the design of a failure detection service for large scale distributed systems. In Proc. Intl. Symp. Towards Peta-Bit Ultra-Networks (PBit 2003), pp. 88-95, Ishikawa, Japan, September 2003.

[Delgrossi95] L. Delgrossi and L. Berger. Internet stream protocol Version 2 (ST2) protocol specification - Version ST2+, Internet Request For Comments, August 1995, RFC 1819.

[Dertouzos99] Dertouzos, M. The future of computing. Scientific American, August 1999. http://www.sciam.com/1999/0899issue/0899dertouzos.html

[Dertouzos99] M. Dertouzos, "The Oxygen Project," Scientific American, Vol. 281 No. 2, 52-63, August 1999

[Dey01] A.K. Dey. Understanding and Using Context. Personal and Ubiquitous Computing, 5(1), 2001.

[DiPippo97] L. DiPippo, R. Ginis, M. Squadrito, S. Wohlever, V. Wolfe, and I. Zykh. Expressing and enforcing Timing Constraints in a CORBA Environment. Technical report TR97-252. University of Rhode Island, 1997.

[Dixit02] Sudhir Dixit and Ramjee Prasad. Wireless IP and Building the Mobile Internet (Artech House Universal Personal Communications Series), Artech House; ISBN: 158053354X, November 2002

[Douence00] Rémi Douence, and Mario Südolt. On the Lightweight and Selective Introduction of Reflective Capabilities in Applications. ECOOP'2000 Workshop on Reflection and Metalevel Architectures, Sophia Antipolis and Cannes, France June 12 - 16, 2000

[Eddon98] Eddon, G. and Eddon, H. Inside Distributed Com, Microsoft Press, 1998

[Eide01] V. Eide and F. Eliassen and O. Lysne. Supporting Distributed Processing of Time-based Media Streams. Distributed Objects and Applications, DOA, 2001

[ElGebaly98] Hani ElGebaly. Characterization of Multimedia Streams of an H.323 Terminal. Intel Technology Journal, April-June 1998

[Elson01] J. Elson, A. Cerpa, ICAP the Internet Content Adaptation Protocol. Internet Draft, The ICAP Protocol Group, http://www.icap.org/spec/icap_specification.txt, June 2001.

[Emblaze03] Emblaze, Inc., http://www.emblaze.com

[Engler98] Engler, D. and Dawson R. The exokernel operating system architecture. Ph.D. thesis, Massachusetts Institute of Technology, October 1998.

[Esler99] M. Esler at al. Next Century Challenges: Data-Centric Networking for Invisible Computing, in Proc. MobiCom '99, 256-262, August 1999

[Ewald02] Ewald, T. Transactional COM+: Building Scalable Applications (AW DevelopMentor Series), Addison Wesley; ISBN: 0201615940, March, 2001

[Fairhurst98] G. Fairhurst. Plenary lecture: Operation, Network Delivery, and Applications of Digital TV using the MPEG-2 Standard, Networkshop 26, Aberdeen, UK, 1998.

[Fankhauser99] G Fankhauser, M Dasen, N Weiler, B Plattner, and B Stiller. WaveVideo - An Integrated Approach to Adaptive Wireless Video. ACM Monet, Special Issue on Adaptive Mobile Networking and Computing, vol 4, no 4, 1999

[Feamster99] N Feamster and S Wee. An MPEG-2 to H263 Transcoder. SPIE Voice Video, and Data Communications Conference, September 1999 Boston, MA

[Fast00] http://siliconvalley.internet.com/news/article.php/461231

[FCE03] FCE: Future Computing Environments, http://www.cc.gatech.edu/fce/

[Feinberg02] Eugene A. Feinberg and Adam Shwartz. Handbook of Markov Decision Processes - Methods and Applications, USA, Kluwer 2002

[Feng99] W. Feng, M. Liu, B. Krishnaswami, and A. Prabhudev. A priority-based technique for the best-effort delivery of stored video. Proceedings of Multimedia Computing and Networking, January 1999.

[Feng01] Feng, Y., Zhu, J., Wireless Java Programming with J2ME, Sams Publishing, 1st Edition, 2001

[Fenner00] B. Fenner, M. Handley, H. Holbrook, I. Kouvelas. Protocol Independent Multicast - Sparse Mode (PIM-SM): Protocol Specification (Revised), Work In Progress, <draft-ietf-pim-sm-v2-new-01.txt>, 2000.

[Ferg98] Ferguson, P and Senie, D. Network ingress Filtering: Defeating Denial of Service Attacks which Employ IP source address spoofing. RFC 2267, 1998

[Ferrari90] Ferrari, D., Client Requirements For Real-Time Communication Services, IEEE Communications, 65-72, Nov 1990

[Ferrari94] D. Ferrari, A. Banerjea, and H. Zhang. Network Support for Multimedia - A Discussion of the Tenet Approach. Computer Networks and ISDN Systems, vol. 26, pp. 1267-1280, 1994.

[Fingar00]   P. Fingar, H. Kumar, and T. Sharma. Enterprise E-Commerce, Published by Meghan-Kiffer Press, Tampa, FL USA. 2000

[Fitz01] Fitz, T. A Webcast Transcoding Gateway. Berkeley Multimedia Research Center, TR 2001-162, December 2001.

[Fiuczynski98] Marc E. Fiuczynski, Richard P. Martin, Brian N. Bershad, David E. Culler.   SPINE: An Operating System for Intelligent Network Adapters. Technical Report UW-CSE-98-08-01, August 1998.

[Floyd93] Sally Floyd and Van Jacobson. Random Early Detection gateways for Congestion Avoidance, IEEE/ACM Transactions on Networking, 1(4):397--413, August 1993.

[Floyd97] S. Floyd, V. Jacobson, C. Liu, S. McCanne, and L. Zhang, "A Reliable Multicast Framework for Light-Weight Session and Application Level Framing," IEEE/ACM Trans. Networking 5, No. 6, Dec 1997.

[Fong03] Bernard Fong, Predrag B. Rapajic, G.Y. Hong, A.C.M. Fong. Factors Causing Uncertainties in Outdoor Wireless Wearable Communications.

[Forouzan02] Forouzan, B. Data Communications and Networking – 2nd Ed. McGraw-Hill Publishers, 2002

[Foster01] I. Foster, C. Kesselman, S. Tuecke. The Anatomy of the Grid: Enabling Scalable Virtual Organizations. International Journal of Supercomputer Applications, 15(3), 2001.

[Fox96] A. Fox, S. D. Gribble, E. Brewer, and E. Amir, Adapting to Network and Client Variability via On-Demand Dynamic Distillation, Proceedings of the Seventh International Conference on Architectural Support for Programming Languages and Operating Systems (ASPLOS), October 1996, pp. 160-170.

[Fritz97] Fritz, Jeffrey. Routing and Switching in ATM Networks: Organizing your ATM network into a hierarchy will make for efficient routing and allow for expansion. Byte Magazine, September 1997

[Fulp01] Errin W. Fulp and Douglas S. Reeves. Optimal Provisioning and Pricing of Differentiated Services Using QoS Class Promotion, Jahrestagung (1), pp. 144-150, 2001

[Furuskar99] A. Furuskar, S. Mazur, F. Muller, and H. Olofsson. EDGE: Enhanced Data Rates for GSM and TDMA/136 evolution. IEEE Personal Communications, Vol. 6, pp. 55--56, June 1999

[Gamma95] Erich Gamma, Richard Helm, Ralph Johnson and John Vlissides. Design Patterns: Elements of reusable object-oriented software. Addison-Wesley, 1995

[Garcia99] J.J. Garcia-Luna-Aceves and E.L. Madruga. A Multicast Routing

Protocol for Ad-Hoc Networks, In Proceedings of IEEE INFO-COM'99, New York, NY, Mar. 1999, pp. 784-792.

[Garg99] Garg, V., Wilkes, J., Principles & Applications of GSM, Prentice Hall, 1999

[Gibson01] Gibson, J. Editor, Multimedia Communications – Directions and innovation. Academic Press in Communications, San Diego, ISBN: 0-12-282160-2, 2001

[Gauthier96] Paul Gauthier, Daishi Harada, and Mark Stemm. Reducing power consumption for the next generation of PDAs: It's in the network interface! http://HTTP.CS.Berkeley.EDU/~stemm, January 1996

[Gokhale02a] Anirudda Gokhale, Balachandran Natarjan, Douglas C. Schmidt, Andrey Nechypurenko, Nanbor Wang, Jeff Gray, Sandeep Neema, Ted Bapty, and Jeff Parsons, CoSMIC: An MDA Generative Tool for Distributed Real-time and Embdedded Component Middleware and Applications, OOPSLA 2002 Workshop on Generative Techniques in the Context of Model Driven Architecture, Seattle, WA, November 2002.

[Gokhale02b] Aniruddha Gokhale, Balachandran Natarajan, Joseph Cross, and Douglas C. Schmidt, Towards Dependable Real-time CORBA Middleware, Cluster Computing: the Journal on Networks, Software, and Applications Special Issue on Dependable Distributed Systems, edited by Alan George, to appear, 2002.

[Golm97a] Golm, Michael and Kleinöder, Jürgen. MetaJava, Presented at STJA '97, September 10, 1997, Erfurt, Germany;

[Golm97b] Golm, M. (1997). Design and Implementation of a Meta Architecture for Java, University of Erlangen, MSc Thesis.

[Golm98] Golm, Michael. metaXa and the Future of Reflection, Presented at the OOPSLA Workshop on Reflective Programming in C++ and Java, October 18, 1998, Vancouver, British Columbia

[Golm99] Michael Golm, Jürgen Kleinöder. Jumping to the Meta Level: Behavioural Reflection can be fast and flexible. Meta-Level Architectures and Reflection Lecture Notes in Computer Science 1616, Springer-Verlag, Berlin, Heidelberg, New York, Tokyo, pp. 22-39, 1999

[Gopalakrishnan99] Gopalakrishnan, S., Reininger, D. and Ott, M. Realtime MPEG system stream transcoder for heterogeneous networks, Proceedings of Packet Video '99, pp. 56-65, Columbia University, New York City, April 1999.

[Gopalakrishnan99] R. Gopalakrishnan, J. Griffioen, G. Hjalmtysson, and C. Sreenan. Stability and fairness issues in layered multicast, in Workshop on Network and Operating System Support for Digital Audio and Video (NOSS-DAV), (Basking Ridge, New Jersey, USA), June 1999.

[Gopalakrishnan00] R. Gopalakrishnan, J. Griffioen, G. Hjalmtysson, C. Sreenan, and S. Wen. A simple loss differentiation approach to layered multicast, in IEEE Infocom, (Tel Aviv, ISRAEL), March 2000.

[Gordon00] Gordon, A, Gordon, A.I. The COM and COM+ Programming Primer. Prentice Hall, March 2000

[Goralski98] Goralski, Walter. ADSL and DSL Technologies. McGraw-Hill Osborne Media; ASIN: 0070246793 March 6, 1998

[Gossain02] H. Gossain and C. Cordeiro and D. Agrawal. Multicast: Wired to Wireless. IEEE Communications Magazine, pp. 116-123, June 2002

[Grant02] Scott Grant, Michael P. Kovacs, Meeraj Kunnumpurath, Silvano Maffeis, K. Scott Morrison, Gopalan Suresh Raj, Paul Giotta, James McGovern. Professional JMS, ISBN: 1861004931, Wrox Press Inc; 2nd edition, June 02

[Greenberg99] Greenberg, S., Boyle, M. & Laberge, J. PDAs and Shared Public Displays: Making Personal Information Public, and Public Information Personal. In: Personal Technologies, Vol.3, No.1, pp. 54-64, 1999

[Gribble01] S. Gribble, et al., The Ninja architecture for robust Internet-scale systems and services. Computer Networks, Vol. 35, No. 4, 2001

[Griwodz99] C Griwodz, M Zink, M Liepert, and G. Steinmetz. Internet VoD Cache Server Design. ACM Multimedia 99, pp. 123-126, October 1999

[Grossman02] D. Grossman. New terminology and clarifications for diffserv, Request for Comments 3260, Internet Engineering Task Force, April 2002

[Hamilton93] G. Hamilton, M. Powell, and J. Mitchell. Subcontract: A Flexible Base for Distributed Programming. Proceedings ACM 14th Symposium on Operating Systems Principles, Nt. Carolina, Dec 1993.

[Hamilton94] Hamilton (G.), Kougioris (P.), The Spring Nucleus: a Micro-kernel for objects, in Proceedings of the Usenix Summer Conference, Ohio, United States, June 1994, pages 147-159.

[Han99] Han, R., Bhagwat, P., LaMarie, R., Mummert, T., Dynamic adaption in an image transcoding proxy for mobile web browsing, IEEE personal communications, Vol. 5, No. 6, pp. 8-17, December 1998,

[Han01] Han, R., Lin, C., Smith, J., Tseng, B. Variable Complexity Compression and Decompression of Multimedia For CPU/Power-Constrained Mobile Devices. ACM Intl. Workshop on Wireless Mobile Multimedia (WOWMOM 2001), Rome, Italy, July 2001

[Handley96] Handley, Mark. SAP: Session Announcement Protocol. November 1996.

[Handley97] Handley, Mark. The SDR Session Directory. Tech Report: University College London. http://mice.ed.ac.uk/mice/archive/sdr.html, 1997

[Handley99] M Handley, H Schulzrinne, E Schooler, and J Rosenberg. Session Initiation Protocol (SIP); IETF RFC 2543, March 1999

[Handley00] Handley, M., Rejaie, R. and Estrin, D. Layered Quality Adaptation for Internet Video Streaming. IEEE Journal on Selected Areas of Communications (JSAC). Special issue on Internet QOS, Winter 2000.

[Hannemann02] Jan Hannemann and Gregor Kiczales. Design Pattern Implementations in Java and AspectJ. In Proc. of OOPSLA, ACM, 2002

[Hansmann03] Uwe Hansmann, Lothar Merk, Martin S. Nicklous, Thomas Stober. Pervasive Computing (Springer Professional Computing Guide). Springer-Verlag Berlin and Heidelberg GmbH & Co. KG; ISBN: 3540002189, June 2003

[Harel86] Harel, D. Statecharts: A visual approach to complex systems (revisited), Report CS86-02, DEP. App Maths Weizman Instutute of Science, Rehovot, Israel, March 1986

[Harrysson02] Anders Harrysson. Industry challanges for mobile services. Fourth IEEE Conference on Mobile and Wireless Communications Networks (MWCN 2002), Stockholm, Sweden, September 9-11, 2002

[Hartenstein01] H Hartenstein, A Schrader, A Kassler, M Krautgaertner, and C Niedenneier. High Quality Mobile Communication, Proc of the KNS 2001, March 2001

[He02] Dan He, Gilles Muller, Julia L. Lawall. Distributing MPEG movies over the Internet using programmable networks. Proceedings of ICDCS 2002 - The 22nd International Conference on Distributed Computing Systems, July 2-5, 2002 - Vienna, Austria

[Hector02] Hector A. Duran-Limon, Gordon S. Blair. Reconfiguration of Resources in Middleware. 7th IEEE International Workshop on Object-oriented Real-time Dependable Systems, San Diego, CA, January, 2002.

[Hemy99] Hemy, M., Hengartner, U., Steenkiste, P. and Gross, T., MPEG system streams in best-effort networks, Proceedings of IEEE packet Video '99, Columbia University, NY City, April 1999

[Heinzelman00] W. Heinzelman. *Application-Specific Protocol Architectures for Wireless Networks.* PhD Thesis, Massachusetts Institute of Technology, June 2000.

[Hof00] Markus A. Hof. Using Reflection for Composable Message Semantics. ECOOP'2000 Workshop on Reflection and Metalevel Architectures, Sophia

Antipolis and Cannes, France June 12 - 16, 2000

[Hoffman96] M. Hoffman, A Generic Concept for large-Scale Multicast, Proc. Int. Zurich Seminar on digital communications (IZS'96), February 1996

[Hoffman02] Hoffman, J. GPRS demystified. McGraw-Hill Professional; 1st edition, ISBN: 0071385533, September 4, 2002

[Hokimoto97] A.Hokimoto, T.Nakajima, "Robust Host Mobility Supports for Adaptive Mobile Applications", In Proceedings of the International Conference on World Wide Computing and its Applications'97, 1997.

[Huang02] L. Huang, U. Horn, F. Hartung und M. Kampmann. Proxy-based TCP-friendly Streaming over Mobile Networks, Proceedings 5th ACM Workshop on Wireless and Mobile Multimedia (WOWMOM 2002), Atlanta, September 2002.

[Hughes01] Hughes, C.J., Srinivasan and Adve, S. Saving energy with architectural and frequency adaptations for multimedia applications. Proceedings of the 34th international symposium on microarchitecture, Dec 2001

[Huni95] Huni, H., Johnson, R. & Engel, R. A framework for network protocol software in Object-Oriented Programming systems, Languages and Applications Conference Proceedings (OOPSLA'95), ACM Press, 1995

[Huitema90] Christian Huitema and Walid Dabbous. "Extension of OSI TP4 to Support Transport Bridging", In Proceedings of Workshop for CL/CO internetworking, Washington, USA, July 1990.

[Hunt98] G. D. H. Hunt, G. S. Goldszmidt, R. P. King, and R. Mukherjee, "Network Dispatcher: A Connection Router for Scalable Internet Services," Computer Networks and ISDN Systems. 30, 347-357 (April 1998).

[Huston00] Huston, G. Next Steps for the IP QoS Architecture. Internet Architecture Board, (draft -iab-qos-00.txt) March 2000.

[Huston01] Huston, G. Internet Performance Survival Guide: QoS Strategies for Multiservice Networks. ISBN 0-471-37808-9, John Wiley & Sons, 2001

[H261] Recommendation H.261: Video codec for audiovisual service at 64 Kbits/s, International Telecommunication Union (ITU-T), 1993

[IBM97] IBM. Bean Extender Documentation, version 2.0 (1997).

[IBM01] AV Project. www.haifa.il.ibm.com/projects/multimedia/audio_video/compression.html

[IETF03] IETF Home Page, http://www.ietf.org/

[Infopad99] INFOPAD. UC Berkeley, http://infopad.eecs.berkeley.edu/.

[ITUH263-97] ITU-T. Draft Recommendation H.263: Video coding for low bit rate communication. 1997.

[ISO14496] ISO/IEC document 14496. Available for viewing at http://www.tnt.uni-hannover.de/project/mpeg/audio/public/

[ISOC03] The Internet Society. http://www.isoc.org/pubs/

[ISWG03] The Integrated services working group home page http://www.ietf.org/html.charters/intserv-charter.html

[ITU-T02] ITU-T. H.26L Standardisation. http://www.tnt.uni-hannover.de/project/veeg/.

[Ives01] Ives, B., Loiacono, E., and Piccoli, G. DSL Provisioning: Redefining Customer Service. Communications of the Association of Information Systems. December, Vol. 7, No. 21, 2001

[Jamjoom01] Hani Jamjoom and Sugih Jamin and Kang Shin. Self-Organizing Network Services, citeseer.nj.nec.com/434665.html

[Jardetzky95] Jardetzky, P., Sreenan, C., Storage and synchronisation for distributed continuous media, Multimedia Systems, 3, No 3, pp 151-161, Sept 1995

[Javasoft02] JavaSoft, Java 2 Enterprise Edition, http://java.sun.com/j2ee/.

[Jee03] G.I. Jee, S.C. Boo, J.H. Choi, H.S. Kim. An Indoor Positioning Using GPS Repeater. Proc. of (Institute of Navigation) ION GPS/GNSS 2003 - Oregon Convention Center, Portland, Oregon, September 9-12, 2003

[Ji03] Ping Ji, Zihui Ge, Jim Kurose, Don Towsley. Applications, Technologies, Architectures and Protocols for Computer Communication archive. Proceedings of the 2003 conference on Applications, technologies, architectures, and protocols for computer communications table of contents, Karlsruhe, Germany, pp.251-262, ISBN: 1-58113-735-4, 2003

[JMF01] Java Media Framework Specification. Sun Microsystems, http://www.javasoft.com, 2001

[Johanson01] Johanson, M. An RTP to HTTP video gateway. In Proceedings of the tenth International world wide web conference, Hong Kong, May 2001

[Jstamp02] JStamp Homepage, http://jstamp.systronix.com/index.htm

[Joesph97] A. D. Joseph, J. A. Tauber, and M. F. Kaashoek. Mobile computing with the Rover toolkit. IEEE Transactions on Computers, 46(3):337-352, March 1997

[Kammann02] J. Kammann, T. Blachnitzky. Split-Proxy Concept for Application Layer Handover in Mobile Communication Systems. Fourth IEEE Conference on Mobile and Wireless Communications Networks (MWCN 2002), Stockholm, Sweden, September 9-11, 2002

[Katz94] Katz, R. Adaptation and Mobility in Wireless Information Systems. IEEE personal communications, Volume 1, pp. 6-17, January 1994

[Karrer01] Roger Karrer and Thomas Gross, Dynamic Handoff of Multimedia Streams. In Proc. NOSSDAV '01, pages 125-133, Danfords on the Sound, Port Jefferson, New York, June 2001

[Kassler00a] A Kassler, A Neubeck, and P Schulthess. Real-time Filtering of Wavelet Coded Videostreams for Meeting QoS Constraints and User Priorities. Proc of Packet Video Workshop, May 2000

[Kassler00b] A Kassler and A Neubeck. Self learning Video Filters for Wavelet Coded Videostreams. Proceedings of the Int. Conf of image Processing (ICIP2000), September 2000

[Kassler01a] A Kassler, A Neubeck, and P Schulthess. Classification and evaluation of Filters for Wavelet Coded Videostreams. Signal Processing Image Communication, vol. 6, pp 795- 807, May 2001

[Kassler01b] A Kassler, C Kucherer, and A Schrader. Adaptive Wavelet Video Filtering, in Proc 2nd International Workshop on Quality of future Internet Services (QofIS), Coimbra, Portugal, September 2001

[Kassler02] A. Kassler. Enabling Mobile Heterogeneous Networking Environments with End-to-End User Perceived QoS - The BRAIN vision and the MIND approach. Proceedings of the European Wireless Conference, pp- 503-509, Florence, Italy February, 2002

[Khansari94] Khansari, M., Jalali, A., Dubois, E., and Mermelstein, P. Robust low bit-rate video transmission over wireless access systems, Proceedings of International Communications Conference , pp. 571-575, 1994.

[Kiczales91] Kiczales, G. / Rivières, J. / Bobrow, D. G. The Art of the Metaobject Protocol, MIT Press, 1991.

[Kiczales96] Gregor Kiczales. Beyond the Black Box: Open Implementation, IEEE Software, January. 1996.

[Kiczales97] Gregor Kiczales, John Lamping, Cristina Videira Lopes, Anurag Mendhekar and Gail Murphy. Open Implementation Design Guidelines, Proceedings of International Conference on Software Engineering, Boston Ma, May 1997.

[Kim93] Kim, Y. and Un, C. Analysis of bandwidth allocation strategies with access restrictions in broadband ISDN. IEEE Transactions on Communications,

Vol. 41, No. 5, pp. 771-781, May 1993

[Kim03] S.J. Kim, G.I. Jee, J.G. Lee. A Method for Repeater Position Determination in the Wireless Location. Proc. of (Institute of Navigation) ION GPS/GNSS 2003 - Oregon Convention Center, Portland, Oregon, September 9-12, 2003

[Kitavaks91] Kitavaks, N., Itah, K. Pure delay effects in speech quality communications, IEEE JSITC, Vol 9, No 4, pp. 586-93, May 1991

[Knudsen00] Knudsen, Ken. Building PerfMon Support into Applications using COM. Microsoft Systems Journal (MSJ), February 2000.

[Kochnev03] Dmitry S. Kochnev, Andrey A. Terekhov. Surviving Java for Mobiles. IEEE Pervasive Computing, (Vol. 2, No. 2), pp. 90-95, April-June 2003

[Koenen02] Koenen, Rob Mpeg-4 Overview - (V.21 – Jeju Version), International Organisation For Standardisation, Iso/Iec Jtc1/Sc29/Wg11 - Coding Of Moving Pictures And Audio Iso/Iec Jtc1/Sc29/Wg11 N4668, March 2002

[Kojo97] Kojo M., Raatikainen K., Liljeberg M., Kiiskinen J., Alanko T.: An Efficient Transport Service for Slow Wireless Telephone Links. IEEE Journal on Selected Areas in Communications, Vol. 15, No. 7, September 1997.

[Kompella93] V.P. Kompella, J.C. Pasquale, and G.C. Polyzos. Multicast routing for multimedia communication' IEEE/ACM Transactions on Networking, vol. 1, no. 3, pp. 286--292, June 1993.

[Kon00] Kon, F., Roman, R. Liu, P, Mao, J. Yamane, T. Campbell, R. Monitoring, Security, and Dynamic Configuration with the DynamicTAO Reflective ORB, Proceedings of the IFIP/ACM International Conference on Distributed Systems Platforms and Open Distributed Processing, Middleware 2000

[Kotz00] Kotz, D., Cybenko, G., Gray, R., Jiang, G. and Ronald, A. Performance Analysis of Mobile Agents for Filtering Data streams on Wireless Networks, MSWiM, Boston USA 2000

[Kouvelas98] Kouvelas, I., Hardman, V., Crowcroft, J. Network Adaptive Continuous Media Application through Self Organised Transcoding. Proceedings of NOSSDAV 1998, http://www.nossdav.org/1998/papers/nossdav98-004.pdf, Cambridge, UK, July 1998

[Kravets01] Kravets, R. Krishnan, P. Application-driven Power Management For Mobile Communications," Wireless Networks, vol. 6, no. 4, pp. 263-277, 2001.

[Krishna03] Arvind Krishna, Douglas C. Schmidt, Krishna Raman, and Raymond Klefstad, Optimizing the ORB Core to Enhance Real-time CORBA Predictability and Performance. 5th International Symposium on Distributed Objects and

Applications (DOA), Catania, Sicily, Nov 2003.

[Kreller98] B. Kreller. UMTS: A Middleware and Mobile-API Approach. IEEE Personal Communications, Vol. 5, No. 2, pp 32-38, April 1998

[Kunz02] Thomas Kunz and Ed Cheng. On-Demand Multicasting in Ad-Hoc Networks: Comparing AODV and ODMRP. Proceedings of ICDCS 2002 - The 22nd International Conference on Distributed Computing Systems, July 2-5, 2002 - Vienna, Austria

[Kyas02] Othmar Kyas, Gregan Crawford. ATM networks. Prentice Hall Publishers, 1st Edition. ISBN: 0130936014 May, 2002

[Lee00] Lee, R. and Nathuji, R. Power and performance analysis of PDA architectures. http://www.cag.lcs.mit.edu/6.893-f2000/project/, December 2000

[Lee00b] Lee, R. Power and Performance Analysis of PDA Architectures, MIT Internal Report at http://www.cag.lcs.mit.edu/6.893-f2000/project/lee_final.pdf, December, 2000

[Leonard96] Leonard, J.N. Franken and B.R. Haverkort. Reconfiguring Distributed Systems using Markov Decision Models. Trends in Distributed Systems (TreDS'96), Aachen, Germany, pp. 219-228, Oct 1996

[Ledoux99] Ledoux, T. OpenCorba: A reflective Open Broker. Proceedings of the 2nd International Conference on Reflection'99, pp. 197-214, Saint Malo, France 1999

[Legout00] A. Legout and W. Biersack, "PLM: Fast convergence for cumulative layered multicast transmission schemes," in *ACM Sigmetrics*, (Santa Clara, Califor-nia, USA), June 2000.

[Lehman98] L. Lehman, G. S., and D. Tennenhouse. Active reliable multicast. In IEEE INFOCOM '98, San Francisco, March 1998.

[Leiner89] Leiner, B., Critical Issues in High Bandwidth Networking, RFC 1077, DARPA, November 1989.

[Li96] Li X. and Ammar M. H., Bandwidth control for replicated-stream multicast video distribution, 5th International Symposium on High Performance Distributed Computing, HPDC '96, pp. 356-363, Syracuse, NY, August 1996.

[Li97] S. Li and B. Bhargava. Active gateway: a facility for video conferencing traffic control. In Proceedings of the 21st Annual International Computer Software and Applications Conference (COMPSAC), 1997.

[Li98] Li, W. Bit plane coding of DCT coefficients for fine granularity scalability, ISO/IEC JTC1/SC29/WG11, MPEG98/M3989, October 1998

[Liebl01] G Liebl et al. An RTP Payload Format for Erasure Resilient

Transmission of Progressive Multimedia Streams. Internet Engineering Task Force, Work in Progress, February 2001

[Liljberg96] LILJEBERG, M., ET AL. Enhanced Services for World Wide Web in Mobile WAN Environments. Tech. Rep. C-1996-28, University of Helsinki CS, Apr. 1996.

[Lin01] Yi-Bing Lin and Imrich Chlamtac. Wireless and Mobile Network Architectures. John Wiley, 2001

[Lippert99]Martin Lippert and Cristina Videira Lopes. A Study on Exception Detection and Handling Using Aspect-Oriented Programming. Xerox PARC Technical Report P9910229 CSL-99-01, 1999

[Lopez98] Cristina Videira Lopes and Gregor Kiczales. Recent Developments in AspectJ. In ECOOP'98 Workshop Reader, Springer-Verlag LNCS 1543, Springer-Verlag, 1998.

[Long96] Long, A. C. Full-motion Video for Portable Multimedia Terminals, A project report submitted in partial satisfaction of the requirements for the degree of Master of Science in Computer Science, University of California, Berkeley, 1996.

[Lorenz03] David H. Lorenz, John Vlissides. Pluggable reflection: decoupling meta-interface and implementation. International Conference on Software Engineering, Proceedings of the 25th international conference on Software engineering, ISSN:0270-5257, Portland, Oregon, pp: 3–13 2003

[Loyall01] Loyall JL, Gossett JM, Gill CD, Schantz RE, Zinky JA, Pal P, Shapiro R, Rodrigues C, Atighetchi M, Karr D. Comparing and Contrasting Adaptive Middleware Support in Wide-Area and Embedded Distributed Object Applications. Proceedings of 21st IEEE International Conference on Distributed Computing Systems (ICDCS-21), April, 2001, Phoenix, AZ.

[Maes87] Maes, P. Concepts and experiments in computational reflection. OOPSLA'97, 1987

[Maffeis96] Maffeis Silvano, Bischofberger Walter, and Mätzel Kai-Uwe. GTS: A Generic Multicast Transport Service to Support Disconnected Operation. ACM Wireless Networks Journal, 2, 1, pp. 87- 96, 1996

[Maffeis97] Maffeis, S. Building Reliable Distributed Systems with CORBA, Theory and Practice of Object Systems, Vol. 3(1), John Wiley & Sons, April 1997

[Maffeis02] Maffeis, S. Mobile Services for Java-enabled Devices on 3G Wireless Networks, In: World market Research Report, 2002 http://www.softwired-inc.com/people/maffeis/publications.html

[Maniatis99] P. Maniatis et al. The Mobile People Architecture. ACM Mobile

Computing and Communications Review, July 1999.

[Marangozova00] Vania Marangozova and Fabienne Boyer. Using Reflective Features to Support Mobile Users. ECOOP'2000 Workshop on Reflection and Metalevel Architectures, Sophia Antipolis and Cannes, France June 12 - 16, 2000

[Margaritidis00] Margaritidis, M and Ployzos, G. MobiWeb: adaptive enabling continuous media applications over wireless links, IEEE conference on third generation wireless communications, San Francisco, June 2000

[Matthur03] Ashwatha Matthur and Padmavathi Mundur. Congestion Adaptive Streaming: An Integrated Approach. DMS'2003 - The 9th International Conference on Distributed Multimedia Systems, Florida International University Miami, Florida, USA, September 24-26, 2003

[Mauve01] Martin Mauve, Volker Hilt, Christoph Kuhmunch and Wolfgang Effelsberg. Toward a common application level protocol for distributed media. IEEE Trans on Multimedia, Vol. 3,No. 1, pp.152-161, 2001

[Maxemchuk01] N. F. Maxemchuk and D. H. Shur. An {Internet} multicast system for the stock market. ACM Transactions on Computer Systems, Vol. 19, No. 3, pp. 384—412, 2001

[McCanne95] McCanne S., and Jacobson V., vic: A Flexible Framework for Packet Video, Proceedings of ACM Multimedia `95, November 1995.

[McCanne96] S. McCanne, V. Jacobson, and M. Vetterli, "Receiver-Driven Layered Multicast," Proceedings of ACM SIGCOMM '96, August 1996, pp. 117-130.

[Meggers98] J. Meggers, et al. A Multimedia Communication Architecture for Hand-held Devices. Proceedings of the 9th IEEE International Symposium on Personal Indoor and Mobile Radio Communications, Boston, September 1998.

[Minoli02] Minoli, D., Johnson, P., and Minoli, E. Ethernet-Based Metro Area Networks: Planning and Providing the Provider Network. McGraw-Hill, 2002

[Mitzel94] Danny J. Mitzel, Deborah Estrin, Scott Shenker, and Lixia Zhang. An architectural comparison of ST-II and RSVP. In Proceedings of the 13th Annual Joint Conference of the IEEE Computer and Communications Societies on Networking for Global Communciation. Volume 2, pages 716-725, Los Alamitos, CA, USA, June 1994. IEEE Computer Society Press.

[Miyazaki01] A Miyazaki. An RTP Payload Format to Enable Multiple Selective Retransmission. Internet Engineering Task Force, Work in Progress, November 2001

[Modiano99] Eytan Modiano. An adaptive algorithm for optimizing the packet size used in wireless ARQ protocols. MIT Lincoln Laboratory, Lexington, MA 02420-9108, USA. Wireless Networks, Vol (5), No 4 1999

[Mohan99] Rakesh Mohan, John R.Smith, Chung-Sheng Li. Adapting Multimedia Internet Content for Universal Access. IEEE Transactions on Multimedia, Vol.1, No.1, March 1999

[Mondads03] Monads: Adaptation Agents for Nomadic Users, http://www.cs.helsinki.fi/research/monads.

[Mosquito03] MosquitoNet: The Mobile Computing Group at Stanford University, http://mosquitonet.stanford.edu/index.html.

[Haefel01] Richard Monson-Haefel. Enterprise JavaBeans (3rd Edition), O'Reilly & Associates; 3rd edition, October 2001

[MPIF03] MPEG-4 Industry Forum. http://www.m4if.org/tutorials.php

[MPLS00] The Multi Protocol Label Switching working group home page http://www.ietf.org/html.charters/mpls-charter.html

[MPLSforum03] QoS Support in MPLS Networks, PDF available at http://www.mplsforum.org/tech/library.shtml#wp, May 2003

[Nagao01] Katashi Nagao, Yoshinari Shirai, and Kevin Squire. Semantic annotation and transcoding: Making web content more accessible. IEEE Multimedia, 8(2):69–81, April-June 2001.

[Nallet00] Nallet, Pierre. OLE DB Consumer Templates: A Programmers Guide. Addison Wesley; ISBN: 0201657929, 18 October, 2000

[Newman98] Mark Newman, Jason Hong. A Look at Power Consumption and Performance
on the 3Com Palm Pilot, UC Berkeley CS252 Spring 1998 available at http://guir.cs.berkeley.edu/projects/p6/finalpaper.html

[Nichols97] Nichols, K., Jacobson, V., and Zhang, L. A Two-bit Differentiated Services Architecture for the Internet. Internet Draft <draft-nichols-diff-svc> at http://www-nrg.ee.lbl.gov/papers/2bitarch.pdf, Nov 1997.

[Nizhegorodov00] Dmitry Nizhegorodov. Jasper: Type Safe Compile-Time Reflection Language Extensions and MOP Based Templates for Java.ECOOP'2000 Workshop on Reflection and Metalevel Architectures, Sophia Antipolis and Cannes, France June 12 - 16, 2000

[Nguyen01] Nguyen, Vietanh. A Survey of Recent Advances in Wireless Systems. Mobile Computing & Disconnected Operation, pp.45-51, May 2001

[Obraczka98] Katia Obraczka. Multicast Transport Protocols: A Survey and Taxonomy. IEEE Communications magazine, vol. 36, No. 1, January 1998

[OMG97] 'Java RMI To Embrace CORBA/IIOP - Sun and OMG Reaffirm Close

Working   Relationship'   Press   release,
http://www.omg.org/news/pr97/rmiiop.html, June 30, 1997

[OMG02a] Object Management Group. Minimum CORBA Specification,
Revision 1.0, August 2002

[OMG02b] Object Management Group. The Common Object Request Broker:
Architecture and Specification, Revision 3.0, July 2002

[OPES] Open Pluggable Edge Services, http://www.ietf-opes.org.

[Oppenheimer98] D. Oppenheimer and M. Welsh. User Customization of Virtual
Network Interfaces with U-Net/SLE, UC Berkeley Tech Report CSD-98-995,
February, 1998.

[Oxygen02] MIT Project Oxygen, http://www.oxygen.lcs.mit.edu/.

[Pailer01] R. Pailer and J.Stadler. A service framework for carrier grade
multimedia services using PARLAY APIs over a SIP system. Proc. Wireless
Multimedia Workshop, Rome, July 2001.

[Pan95] Pan, Davis. Tutorial on MPEG/Audio Compression. IEEE Multimedia,
Vol. 2, No. 2, pp. 60-74, 1995

[Parlavantzas00] Nikos Parlavantzas, Geoff Coulson, Mike Clarke, and Gordon
Blair. Towards a Reflective Component Based Middleware Architecture.

[Parley03]   The   Parlay   Group,   "PARLAY   specification   2.1",
http://www.parlay.org

[Parr-Curran98a] Parr, G., Curran, K. Optimal Multimedia Transport On The
Internet. Journal of Network and Computer Applications, Vol. 21, pp 149-161,
1998

[Parr-Curran98b] Parr, G., Curran, K. Dynamic Multimedia Protocol Stacks.
Software Concepts and Tools, Vol. 19(2), pp 79-86, 1998

[Parr-Curran98c] Parr, G and Curran, K., Reliable Remote Isochronous
Applications Over The Internet. Presented at IEEE International Conference on
Systems, Man, and Cybernetics, San Diego - CA, October 11-14, 1998

[Parr-Curran99] Parr, G., Curran, K. Multiple Multicast Groups For Multimedia
On The Internet, Information & Software Technology, Vol. 41, pp 91-99, 1999.

[Parr-Curran00] Parr, G., Curran, K. A Paradigm Shift In The Distribution Of
Multimedia. Communications Of The ACM, Vol 43, No 6, pp 103-109, June
2000

[Parr00] Parr, G., Kong, I., Chieng Heng Tze, D., Marshall, A. 'A Mobile Agent
Brokering Environment for The Future Open Network Marketplace'. The Seventh

International Conference on Intelligence in Services and Networks (ISN2000) Greece, ISBN: 3-540-67152-8, page 3-15, Feb. 2000.

[Penner00] Robin R. Penner and Erik S. Steinmetz. Model-Based Design Automation for Mixed-Initiative Interactions with Complex Digital Control Systems. to IEEE Transactions on Systems, Man and Cybernetics, Special issue on Model-Based Cognitive Engineering, April 2000

[Perkins98] Mobile IP - Design Principles and Practices, 1st edition Charles E. Perkins, 0-201-63469-4, Addison-Wesley, 1999

[Perkins03] Perkins, Colin. RTP – Audio and Video for the Internet. ISBN – 0672-32249-8, Pearson Education, June 2003

[Peterson91] Peterson, L. & Hutchinson, N. The X-Kernal: An architecture for implementing network protocols, IEEE Transactions on Software Engineering, 17 (1), pp. 64-76 1991

[Peterson96a] L. L. Peterson. Network Systems Research Group, Department of Computer Science, University of Arizona. x-kernel Programmer's Manual (Version 3.3), Jan. 1996.

[Peterson96b] L. L. Peterson. Getting started with the x-kernel. Network Systems Research Group, Department of Computer Science, University of Arizona, Jan 1996.

[Petrovski03] I. Petrovski, K. Okano, S. Kawaguchi, H. Torimoto, K. Suzuki, M. Toda, J. Akita. Indoor Code and Carrier Phase Positioning with Pseudolites and Multiple GPS Repeaters. Proc. of (Institute of Navigation) ION GPS/GNSS 2003 - Oregon Convention Center, Portland, Oregon, September 9-12, 2003

[Phan02] Thomas Phan, George Zorpas, and Rajive Bagrodia. An Extensible and Scalable Content Adaptation Pipeline Architecture to Support Heterogeneous Clients. Proceedings of ICDCS 2002 - The 22nd International Conference on Distributed Computing Systems, July 2-5, Vienna, Austria, 2002

[Plagemann93] Plagemann, T., Plattner, B., Vogt, M., Walter, T.: Modules as Building Blocks for Protocol Configuration, Proceedings of International Conference on Network Protocols ICNP'93, San Francisco, October 1993, pp. 106-113

[Plagemann95] Plagemann, T., Saethre, K. A., Goebel, V.: "Application Requirements and QoS Negotiation in Multimedia Systems", in: Proceedings of Second Workshop on Protocols for Multimedia Systems, Salzburg Austria, October 1995

[Poger97] Elliot Poger and Mary Baker, Secure Public Internet Access Handler (SPINACH). Proceedings of the USENIX Symposium on Internet Technologies and Systems, December 1997.

[Portolano02] Portolano: An Expedition into Invisible Computing, http://portolano.cs.washington.edu/.

[Pranata02] Pranata, Anthony. Development of Network Service Infrastructure for transcoding multimedia streams. MSc Thesis Nr. 1978, University of Stuttgart, Faculty of Computer Science, May 2002

[Puder00] A. Puder, K. Roemer: MICO: An Open Source CORBA Implementation. 3rd edition, Morgan Kaufmann Publishers, March 2000

[Pulse99] Author(s) anon. PulseOnline: Industry Trends: Industry effort seeks to provide quality of service in IP networks. December 1999

[PV03] Packet Video, http://www.packetvideo.com

[Pyarali02] Irfan Pyarali, Douglas C. Schmidt, and Ron Cytron. Achieving End-to-End Predictability of the TAO Real-time CORBA ORB. Proceedings of the 8th IEEE Real-Time Technology and Applications Symposium, San Jose, CA, September 2002.

[Quicktime03] http://developer.apple.com/quicktime/qtjava/

[Raatikainen99] K. Raatikainen et al. Monads – Adaptation Agents for Nomadic Users. Proc. ITU Telecom'99, October 1999.

[Raatikainen03] Raatikainen, K. Wireless Access & Terminal Mobility in CORBA - version 1.0, Document -- dtc/02-09-16 (Telecom Wireless FTF IDL). The document is available at http://www.omg.org/technology/documents/formal/telecom_wireless.htm

[Rao91] Rao, R. Implementational Reflection in Silica. ECOOP'91 Proceedings, Lecture Notes in Computer Science, P. America (Ed.), Springer-Verlag, 1991.

[Rao01] Rao, H., Chen, Chang, D. iMobile: A proxy based platform for mobile services. Proc. Of the first ACM Workshop on Wireless Mobile Internet (WMI 2001), July 2001

[Rashid89] R. Rashid, D. Julin, D. Orr, R. Sanzi, R. Baron, A. Forin, D. Golub, M. Jones: Mach: A System Software kernel. Proceedings of the 34th Computer Society International Conference (COMPCON 89), February 1989

[Raspall01] F Raspall, C Kubmuench, A Banchs, F Pelizza, and S Sallent. Study of Packet Dropping Policies on Layered Video. Proc of the Workshop on Pocket Video, Korea, 2001

[Redmond00] Barry Redmond, and Vinny Cahill. Iguana/J: Towards a Dynamic and Efficient Reflective Architecture for Java. ECOOP'2000 Workshop on Reflection and Metalevel Architectures, Sophia Antipolis and Cannes, France June 12 - 16, 2000

[Rejaie99] Rejaie, R., Handley, M. and Estrin, D. Quality Adaptation for Unicast Audio and Video. In Proc. ACM SIGCOMM, September 1999.

[Rejaie00] Rejaie, R., Yu, H., Handley, M. and Estrin, D. Multimedia proxy caching mechanism for quality adaptive streaming applications in the Internet. Proceedings of the conference on computer communications. Proc. of IEEE INFOCOM 2000, Tel-Aviv, Israel, March 2000

[Real02] http://www.real.com

[Renesse96] Robbert van Renesse, Ken Birman, and Silvano Maffeis. Horus: A Flexible Group Communication System. Communications of the ACM 39(4) (April 1996).

[Resnick96] Resnick, Ron. Toward the Integration of WWW and Distributed Object Technology: Distributed Objects on the WWW. OPSLA'96 1996

[Resnick99] Resnick, Ron. From JSDA to MASH to SOGS, http://www.infospheres.caltech.edu/mailing-lists/dist-obj/0397.html, July 1999

[RFC761] RFC 761: Transmission Control Protocol (TCP). Available at http://www.ietf.org/rfc/rfc0761.txt?number=761

[RFC768] RFC 768: User Datagram Protocol (UDP). Available at http://www.ietf.org/rfc/rfc0768.txt?number=768

[RFC791] RFC 791: Internet Protocol (IP). http://www.ietf.org/rfc/rfc0791.txt?number=791

[RFC1112] RFC 1112: S.E. Deering, Host extensions for IP multicasting, Aug 1989.

[RFC1889] RFC 1889: Real-Time Protocol (RTP): A Transport Protocol for Real-Time Applications. Available at http://www.ietf.org/rfc/rfc1889.txt?number=1889

[RFC1890] RFC 1890: RTP Profile for Audio and Video Conferences with Minimal Control. Available at http://www.ietf.org/rfc/rfc1890.txt?number=1890

[RFC2068-97] RFC 2068: Internet Engineering Task Force. HyperText Transfer Protocol - HTTP 1.1, Mar. 1997. RFC-2068.

[RFC2205] RFC 2205: Resource ReSerVation Protocol (RSVP)- Version 1 Functional Specification, http://www.ietf.org/rfc/rfc2205.txt

[RFC2210] RFC 2210: The Use of RSVP with IETF Integrated Services, http://www.faqs.org/rfcs/rfc2210.html

[RFC2326] RFC 2326: Real Time Streaming Protocol (RTSP). Available at http://www.ietf.org/rfc/rfc2326.txt?number=2326

[RFC2710] S. Deering, W. Fenner, B. Haberman, "Multicast Listener Discovery (MLD) for IPv6", RFC 2710.

[RFC3376] B. Cain, S Deering, W. Fenner, I Kouvelas, A. Thyagarajan, Internet Group Management Protocol, Version 3, RFC 3376.

[Richardson02] Iain Richardson. Video Codec Design: Developing Image and Video Compression Systems. ISBN: 0471485535, John Wiley & Sons; May, 2002

[Riley98] Sean Riley, Robert Breyer. Switched, Fast, and Gigabit Ethernet (3rd Edition), Que; 3rd Edition, ISBN: 1578700736, December 1998

[Ritchie90] STREAMS Modules and Drivers, UNIX System V Release 4.2, UNIX Press, ISBN 0-13-066879-6

[Rizzo98] L. Rizzo and L. Vicisano. RMDP: An FEC-based Reliable Multicast Protocol for Wireless Environments. Mobile Computing and Communications. Volume 2, Number 2, April 1998.

[Roeder03] Konrad Roeder and Frank D., Jr. Ohrtman. Wifi Handbook : Building 802.11B Wireless Networks. McGraw Hill Text ; ISBN: 0071412514 , March 2003

[Roman00] M. Roman and R.H. Campbell, Gaia: Enabling Active Spaces. Proc. ACM SIGOPS European Workshop, Kolding, Denmark, September 2000.

[Roman01a] Roman, M., Singhai, A., Carvalho, D., Hess, C., and Campbell, R. Integrating PDAs into Distributed Systems: 2K and PalmORB. Proceedings of the International Symposium on Handheld and Ubiquitous Computing (HUC'99), Karlsruhe, Germany, Sept 27-29 1999

[Roman01b] Roman, M., Kon, F. and Campbell, R. Reflective Middleware: From your desk to your hand. IEEE Distributed Systems online Journal. Special issue on Reflective Middleware, July 2001

[Román02] Manuel Román, Christopher Hess, Renato Cerqueira, Anand Ranganathan, Roy H. Campbell, Klara Nahrstedt. A Middleware Infrastructure for Active Spaces. IEEE Pervasive Computing, (Vol. 1, No. 4), pp. 74-83, Oct-Nov 2002

[Rosenberg99] L Rosenberg et al. An RTP Payload Format for Generic Forward Error Correction: Internet Engineering Task Force RFC 2733, December 1999

[Rosenberg02] Rosenberg, J., Schulzrinne, H., Camarillo, G., Johnston, A., Peterson, J., Sparks, R., Handley M., and Schooler, E. SIP: Session Initiation protocol, RFC 3261, Internet Engineering Task Force, June 2002

[Rosenberger98] Rosenberger, J. Teach Yourself CORBA in 14 Days. Sams

Publishing; ISBN: 0672312085, January 1998

[Ross03] Ross, J. The Book of Wi-Fi: Install, Configure, and Use 802.11B Wireless Networking. No Starch Press; ISBN: 188641145X , February 2003

[Roussopoulos03] Mema Roussopoulos and Mary Baker, CUP: Controlled Update Propagation in Peer-to-Peer Networks. Proceedings of the 2003 USENIX Annual Technical Conference, San Antonio, Texas, June, 2003

[Rysavy99] Rysavy, P. The evolution of cellular data: On the road to 3g. http://www.gsmdata.com/today_papers.htm, 1999.

[Saber03] Mahmoud Saber, Nikolay Mirenkov. A Multimedia Programming Environment for Cellular Automata Systems. DMS'2003 - The 9th International Conference on Distributed Multimedia Systems, Florida International University Miami, Florida, USA, September 24-26, 2003

[Sakharov00a] Sakharov, A. A hybrid state machine notation for component specification', SIGPLAN Notices, April 2000.

[Sakharov00b] Sakharov, A. State Machine Specification Directly in Java and C++, Object-Oriented Programming, Systems, Languages, and Applications, OOPSLA 2000, pp. 143-151, Orlando 2000

[Sari99] Sari, H. Broadband Radio Access to Homes and Businesses: MMDS and LMDS. Computer Networks, February, 31(4), 379-393, 1999

[Satyanarayanan96] M. Satyanarayanan. Mobile Information Access. IEEE Personal Communications, Vol. 3 No. 1, February 1996

[Schaphorst96] Richard Schaphorst. Videoconferencing and Videotelephony. Technology and Standards. Artech House, Inc., 685 Canton st., Norwood, MA 02062, 1996.

[Schantz01] Schantz R., Schmidt D. Middleware for Distributed Systems: Evolving the Common Structure for Network-centric Applications, Encyclopaedia of Software Engineering, Wiley & Sons, 2001.

[Schmidt99] Schmidt, D. et al. The design of the TAO real-time object request broker, Computer Communications, Vol 21 (4), pp. 294-324, 1998

[Schmidt01] Schmidt, D et al. Minimum TAO http://www.cs.wustl.edu/~schmidt/ACE_wrappers/docs/mimumumTAO.html

[Schmidt02] Schmidt, D. R&D Advances in Middleware for Distributed, Real-time & Embedded Systems. Communications of the ACM special issue on Middleware, vol. 45, no. 6, June 2002

[Schmidt03] Sven Schmidt, Andreas Marcz, Wolfgang Lehner, Maciej Suchomski, and Klaus Meyer-Wegener. Quality-of-Service Based Delivery of

Multimedia Database Objects without Compromising Format Independence. DMS'2003 - The 9th International Conference on Distributed Multimedia Systems, Florida International University Miami, Florida, USA, September 24-26, 2003

[Schulzr96a] H. Schulzrinne, S. Casner, R. Frederick, V. Jacobson, RTP: A Transport Protocol for Real-Time Applications, http://ds.internic.net/rfc/rfc1889.txt, 1996.

[Schulzr96b] H. Schulzrinne, RTP Profile for Audio and Video Conferences with Minimal Control, http://ds.internic.net/rfc/rfc1890.txt, 1996.

[Scott97] D. Scott Alexander, M. Shaw, S. M. Nettles, and J. M. Smith, Active Bridging, Proceedings of the ACM SIGCOMM '97 Conference on Applications, Technologies, Architectures, and Protocols for Computer Communication, 1997, pp. 101-111.

[SCTE00] Society Of Cable Telecommunications Engineers, Inc. Engineering Committee. Audio codec requirements for the provision of bi-directional audio service over cable television networks using cable modems. Data Standards Subcommittee Document: SCTE DSS-00-01 Date of Original Issue: March 1, 2000 Date of Latest Revision: December 15, 2000

[Seifert98] Rich Seifert. Gigabit Ethernet: Technology and Applications for High-Speed LANs Addison-Wesley Pub Co; 1st edition, ISBN: 0201185539, April 21, 1998

[Senda99] Senda, Yuzo, Harasaki, Hidenobu. A real-time software MPEG transcoder using a novel motion vector reuse and a SIMD optimisation, Proceedings of ICASSP '99, pp. 145-152, Sept 1999

[Shanableh00] T. Shanableh and M. Ghanbari. Heterogeneous video transcoding to lower spatio-temporal resolutions and different encoding formats. IEEE Trans. Multimedia, vol. 2, pp. 101–110, June 2000.

[SIP] Session Initiation Protocol, IETF SIP Working Group, http://www.ietf.org/html.charters/sip-charter.html

[Sisalem98] D. Sisalem and H. Schulzrinne, The Multimedia Internet Terminal, Journal on Telecommunication Systems, Vol. 9, No. 3, pages 423-444, 1998.

[Smith98] Smith, J.R.; Mohan, R.; Chung-Sheng Li. Transcoding internet content for heterogeneous client devices. Proceedings of the 1998 IEEE International Symposium on Circuits and Systems, Vol.3, pp. 599 -602, 1998

[Smith01] Smith, C., Collins, D. 3G Wireless Networks. Barnes and Noble, September 2001

[Snoeren01] Snoeren, Alex., Balakrishnan, Hari, and Kaaeshoek, Frans. Reconsidering IP Mobility. In proceedings of the 8th IEEE workshop on Hot

Topics in Operating Systems (HoTOS-VIII), Schloss Elmau, Germany, May 2001.

[Solon03] Tony Solon, Paul McKevitt and Kevin Curran. Telemorph - Bandwidth determined mobile multimodal presentation. IT&T 2003 - Information Technology and Telecommunications
Letterkenny Institute of Technology, Co. Donegal, Ireland. 22-23rd October, 2003

[Srivastava97] Srivastava, Mishra. On quality of service in mobile wireless networks. Proceedings of 7th International Workshop on Network and Operating System Support for Digital Audio and Video (NOSSDAV '97), IEEE, pp 147-58, 19-21 May 1997

[Stallings98] Stallings, William. ISDN and Broadband ISDN with Frame Relay and ATM (4th Edition). Prentice Hall; 4th edition, ISBN: 0139737448, 1998

[Stanford03] Vince Stanford. Pervasive Computing Goes the Last 100 Feet with RFID Systems. IEEE Pervasive Computing, (Vol. 2, No. 2), pp. 9-14, April-June 2003

[Starr99] Starr, T., and Cioffi, J. M., and Silverman, P. J. Understanding Digital Subscriber Line Technology. Prentice-Hall, 1999

[Stefanov00] Todor Stefanov, Paul Lieverse, Ed Deprettere and Pieter van der Wolf. Y-Chart Based System Level Performance Analysis: An M-JPEG Case Study, Proceedings of the Progress Workshop, pp.134-46, October 2000.

[Steinmetz92] Steinmetz, R., and Meyer, T., Modelling distributed multimedia applications, In: International Workshop on Advanced Communications and Applications for High Speed Networks, Munich, Germany, March 16-19, 1992, 337-349.

[Steinmetz96] R. Steinmetz and Klara Nahrstedt. Multimedia: Computing, Communications and Applications. Prentice Hall Inc., NJ, USA, 1996

[Stemm96a] Mark Stemm, Paul Gauthier, Daishi Harada, and Randy H. Katz. Reducing power consumption of network interfaces in hand-held devices. In 3rd International Workshop on Mobile Multimedia Communications (MoMuc-3), Princeton, New Jersey, 1996.

[Steyaert98] Steyaert, Patrick. Open Design of Object-Oriented Languages: A foundation for specialisable reflective language frameworks. PhD thesis: Vancouver University, 1998.

[Striegel02] Aaron Striegel. Scalable approaches for DiffServ Multicasting. Ph.D. Dissertation, November 2002

[Stuckmann02] P. Stuckmann. The GSM Evolution - Mobile Packet Data Services. John Wiley and Sons Ltd; ISBN: 0470848553, September 2002

173

[Sullivan01] Sullivan, G.T. Aspect-oriented programming using reflection and meta-object protocols. Communications of the ACM, Vol. 44, No. 10, pp. 95-97, October 2001.

[Sybex02] Sybex. Networking Complete. Sybex International, ISBN: 0782141439, 26 September, 2002

[Tai02] Stefan Tai, Thomas Mikalsen, Isabelle Rouvellou, Stanley M. Sutton Jr. Conditional Messaging: Extending Reliable Messaging with Application Conditions, Proceedings of ICDCS 2002 – the 22nd International Conference on Distributed Computing Systems, Vienna, Austria, July 2-5 2002

[Talley97] Talley, Terry M., A transmission control framework for continuous media. PhD Thesis: University of North Carolina at Chapel Hill, 1997.

[Tan01] W. Tan and A. Zakhor. Packet classication schemes for streaming MPEG video over delay and loss differentiated networks. Proceedings of the 11th International Packet Video Workshop, Kyongju, Korea, 2001

[Tanter02] Tanter, Eric., Vernaillen, Michael, Piquer, Jose. Towards transparent adaptation of migration policies. Position paper submitted to EWMOS 2002, Chile, pp. 34-39, 2002

[Tanenbaum91] A.S. Tanenbaum, M.F. Kaashoek, R. van Renesse, and H. Bal: The Amoeba Distributed Operating System-A Status Report. Computer Communications, vol. 14, pp. 324-335, July/August 1991

[Tanenbaum03] Tanenbaum, Andrew. Computer Networks, Fourth Edition, ISBN: 0-13-066102-3, Prentice Hall, May 2003

[Tanter03] Tanter, E., P. Ebraert. A flexible approach to run-time inspection. ASARTI workshop at ECOOP, pp. 1-18, July 2003

[Tassel97] Jérôme Tassel, Bob Briscoe, Alan Smith. An End to End Price-Based QoS Control Component Using Reflective Java, in Lecture Notes in Computer Science from the 4th COST237 workshop (Springer-Verlag), December 1997.

[Tatsubori98] Tatsubori, Michiaki and Chiba, Shigeru. Programming Support of Design Patterns with Compile-time Reflection. OOPSLA'98 Workshop on Reflective Programming in C++ and Java, pp.56-60, ISSN 1344-3135, Vancouver,

[Tatsubori99] Tatsubori , Michiaki. An Extension Mechanism for the Java Language. Master of Engineering Dissertation, Graduate School of Engineering, University of Tsukuba , University of Tsukuba, Ibaraki, Japan, Feb 2, 1999.

[Tatsubori00] Michiaki Tatsubori, Shigeru Chiba, Marc-Olivier Killijian and Kozo Itano. OpenJava: A Class-Based Macro System for Java. Lecture Notes in Computer Science 1826, Reflection and Software Engineering, Springer-Verlag,

174

pp.117-133, 2000.

[Tatsubori01] Michiaki Tatsubori, Toshiyuki Sasaki, Shigeru Chiba and Kozo Itano. A Bytecode Translator for Distributed Execution of "Legacy" Java Software. ECOOP 2001 -- Object-Oriented Programming, LNCS 2072, Springer Verlag, pp.236-255, 2001.

[Templeman03] Templeman, J. COM Programming with .NET. Microsoft Press International, ISBN: 0735618755, 2003

[Tennen96] D. L. Tennenhouse and D. J. Wetherall, Towards an Active Network Architecture, Computer Commun. Rev. (originally published in Multimedia Computing and Networking '96) 26, No. 2, 5-18 April 1996

[Tennen97] D. L. Tennenhouse, J. M. Smith, W. D. Sincoskie, D. J. Wetherall, and G. J. Minden, A Survey of Active Network Research, IEEE Communications Magazine 35, No. 1, 80-86 (January 1997).

[Terry94] Terry D. Session Guarantees for Weakly Consistent Replicated Data, Proceedings International Conference on Parallel and Distributed Information Systems, Austin, Texas, September 1994, pp.140-149.

[Terzis99] A Terzis, L Wang, J Ogawa, and L- Zhang. A two-tier Resource management model for the internet. Proceedings of Global Internet, December 1999

[Terzis00] Terzis, A., Krawczyk, J., Wroclawski, J., Zhang, L., RSVP Operation Over IP Tunnels, RFC 2746, January 2000.

[Teitlebaum98] Teitlebaum,B. QoS Requirements for Internet2. First Internet2 Joint Applications and Engineering QoS Workshop, Santa Clara, CA May 21-22, 1998

[Thai99] Thai, Thuan L. Learning DCOM, O'Reilly UK, April 1999

[Thai00] Thai B, Senevirante A. The Use of Software Agents as Proxies. Proceedings of 5th IEEE Symposium on Computers and Communications (ISCC 2000), Antibes, France, p.546-551, 2000

[Thorson01] Thorson, M. VIC viewer for Pocket PC. Available online at http://www.oncoursetech.com/video/default.htm, April 2001

[Todd99] Todd D. Hodes, Steven E. Czerwinski, Ben Y. Zhao, Anthony D. Joseph, Randy H. Katz, An Architecture for a Secure, Wide-Area Service Discovery Service. Wireless Networks Journal (ACM and Baltzer Science Publishers), Special issue on selected papers from MobiCom 99.

[Tojo03] Takuya Tojo, Tomoya Enokido, and Makoto Takizawa. Notification-Based QoS Control Protocol for Group Communication. DMS'2003 - The 9th International Conference on Distributed Multimedia Systems, Florida

International University Miami, Florida, USA, September 24-26, 2003

[Tramontana00] Emiliano Tramontana. Reflective Architecture for Changing Objects. ECOOP'2000 Workshop on Reflection and Metalevel Architectures, Sophia Antipolis and Cannes, France June 12 - 16, 2000

[Tran02] Duc Tran and Kien Hua and Simon Sheu. A New Caching Architecture for Efficient Video Services on the Internet. IEEE Symposium on Applications and the Internet, June 2002

[Trimintzios01] P. Trimintzios A management and control architecture for providing IP differentiated services in MPLS-based networks, IEEE Communications Magazine, May 2001, pp80-88.

[Turletti96] T. Turletti and C. Huitema, Videoconferencing on the Internet, IEEE/ACM Trans. Networking, Jun. 1996, 340--351.

[Ubik03] Ubik, S. QoS in layer 2 networks with Cisco Catalyst 3550. CESNet Technical Report 3/2003, Document online at http://www.cesnet.cz/doc/techzpravy/2003/l2qos/l2qos.pdf, April 2003

[Unix90] Streams programmer's guide. Unix System V Release 4.

[Upadhyaya02] Shambhu Upadhyaya, Abhijit Chaudhury, Kevin Kwiat, Mark Weiser. Mobile Computing: Implementing Pervasive Information and Communications Technologies (Operations Research/computer Science Interfaces), Kluwer Academic Publishers; ISBN:140207137X, July 2002

[Vacca02] Vacca, John. I-mode crash course. Mc-Graw Hill, 2002

[Valenzuela02] S. Gonza lez-Valenzuela, V.C.M. Leung, QoS routing for MPLS networks employing mobile agents, IEEE Network, No. 6, Vol. 3, 2002

[Vandalore99] Vandalore, et al. AQuaFWiN: adaptive QoS framework for multimedia in wireless networks and its comparison with other QoS frameworks. Proceedings 24th Conference on Local Computer Networks. LCN'99, IEEE Computer Soc, pp 88-97, 18-20 Oct. 1999

[VanSteen98] Van Steen, M., Tanenbaum, A.S., Kuz, I and Sips, H.J. A scalable middleware solution for advanced wide-area web services. In Proceedings of Middleware '98 (IFIP International Conference on Distributed Systems Platforms and Open Distributed Processing), pages 37-53, Springer-Verlag, Sept 1998

[VanSteen99] Van Steen, M., Tanenbaum and Homberg, P. Globe: A wide-area distributed system. IEEE Concurrency, pages 70-78, January-March 1999

[Vayssière00] Julien Vayssière. Security and Meta Programming in Java. ECOOP'2000 Workshop on Reflection and Metalevel Architectures, Sophia Antipolis and Cannes, France June 12 - 16, 2000

[Venditto96] Venditto, G., (1996) "Instant video." Internet World, vol. 7, no. 11, November, p.84(11) in Computer ASAP, Article # A18793239, August, 1997.

[Vetro01] A. Vetro, H. Sun and Y. Wang. Object-based transcoding for adaptable video content delivery. IEEE Trans. Circuits and Syst. for Video Tech., March 2001.

[Vinoski97] Vinoski, Steve. Integrating Diverse Applications Within Distributed Heterogeneous Environments, IEEE Communications Magazine, Vol. 14, No. 2, February, 1997.

[Vranyecz98] Z. Vranyecz, K. Fazekas, I. Erenyi, K. Csányi, Comparison Between H263 Coding Method and MPEG-4 in Point of View of VLBR Coding. Proceedings of 5th International Workshop on Systems, Signals and Image Processing (IWSSIP'98), Zagreb, Croatia, June 3-5, 1998

[Queloz99] P.A. Queloz and A. Villazón. Composition of Services with Mobile Code. Proceedings of ASA/MA 99. First International Symposium on Agent Systems and Applications and Third International Symposium on Mobile Agents. Palm Springs, California USA Oct. 1999.

[Villazón00] A. Villazón. A Reflective Active Network Node. Proceedings of the Second International Working Conference on Active Networks (IWAN 2000), Tokyo Japan, Oct. 2000

[W3C03] W3C Home Page, http://www.w3.org/

[Wallace91] Wallace, G. The JPEG Sill Picture Compression Standard, Communications of the ACM, Volume 34, Number 4, pp. 30-44, April 1991

[Wang00] H.J. Wang et al. An Internet-core Network Architecture for Integrated Communications, IEEE Personal Communications, August 2000

[WAP02] WAP Forum. Wireless Application Protocol: White Paper. Available at http://www.wapforum.org/what/WAP_white_pages.pdf, June 2000.

[Warfield02] Andrew Warfield and Yvonne Coady and Norm Hutchinson. Identifying Open Problems in Distributed Systems. Paper online at http://citeseer.nj.nec.com/442270.html, 2002

[Watanabe00] Takuo Watanabe, Amano Noriki, and Kenji Shinbori. A Reflective Framework for Reliable Mobile Agent Systems. ECOOP'2000 Workshop on Reflection and Metalevel Architectures, Sophia Antipolis and Cannes, France June 12 - 16, 2000

[Watson94] Terri Watson. Application design for wireless computing. Proceedings of the IEEE Workshop on Mobile Computing Systems and Applications, Nov. 1994.

[Weiser93] Weiser, M. Computer Science Problems in Ubiquitous Computing.

Comms of the ACM, July 1993.

[Welch98] Welch, Ian and Stroud, R. Using metaobject protocols to adapt third party components. Middleware '98, Lancaster, 1998

[Widmer00] Widmer, Jorg. A mobile network architecture for simulation. International Computer Science Institute, Berkeley, CA. Report: TR-00-009, May 2000.

[Williamson00] Williamson, B. Developing IP Multicast Networks. Indianapolis: Cisco Press, 2000.

[Willebeek96] M. H. Willebeek-LeMair, K. G. Kumar and E. C. Snible. Bamba-- Audio and video streaming over the Internet, Vol 42, No. 2 - Multimedia Systems, 1996

[Willmott99] S. Willmott and B. Faltings. Active organisation for routing. In S. Covaci, editor, Active Networks, Proceedings of the First International Working Conference, Berlin, LNCS 1653, June 1999.

[Wolcott01] Wolcott, P., Press, L., McHenry, W., Goodman, S., and Foster, W. A Framework for Assessing the Global Information Systems, November, 2(6), 2001

[Wong01] Wong, G, Hiltunen, M., and Schlichting, R. A configurable and extensible transport protocol. IEEE Infocom 2001, April 22-26 2001, Anchorage, Alaska, April 2001

[Wrox03] Professional ASP.NET Web Services: Asynchronous Programming by Wrox Press. Article available at http://www.stardeveloper.com/articles/display.html?article=2001121901&page= 1

[Wu96] Wu, Z. and S. Schwiderski (1996). Reflective Java: The Design, Implementation and Applications, Presentation at APM Ltd.

[Wu97] L. Wu, R. Sharma, and B. Smith. Thin streams: An architecture for multicasting layered video. Workshop on Network and Operating System Support for Digital Audio and Video, May 1997.

[Wu02] Wei Wu and Yong Ren and Xiuming Shan. Providing Proportional Loss Rate for Adaptive Traffic: A New Relative DiffServ Model. citeseer.nj.nec.com/wu02providing.html

[Xu97] Xu, Myers, Zhang, and Yavatkar, Resilient Multicast Support for Continuous-Media Applications, NOSSDAV 1997.

[Yamamoto97] Yamamoto, M., Kurose, J., Towsley and Ikeda, H. A delay analysis of sender-initiated and receiver-initiated reliable multicast protocols, proceedings of IEEE INFOCOM'97, pp.408-488, April 1997.

[Yamamoto01] Yamamoto, M., Kurose, J., Towsley and Ikeda, H. "Performance evaluation of ack-based and nak-based flow control schemes for reliable multicase,", proceedings of IEICE Trans. Communications, Vol. E84-B, No. 8, August 2001

[Yates90] Yates, R., Masson, J., Mahe, N. Fiber Optics and Catv Business Strategy. Artech House; ISBN: 089006413X , May 1990

[Yeadon96] N. Yeadon, F. Garcia, D. Hutchinson, and D. Shepherd. Continuous Media Filters for Heterogeneous Internetworking, Multimedia Computing and Networking (MMCN'96), Proc. SPIE 2667, 246-257, January 1996.

[Yoshimura02] Takeshi Yoshimura, Yoshifumi Yonemoto, Tomoyuki Ohya, Minoru Etoh, and Susie Wee. Mobile Streaming Media CDN Enabled by Dynamic SMIL. WWW2002, May 7-11, 2002, Honolulu, Hawaii, USA.

[Yu01] F. Yu, et al., Network-Adaptive Cache Management Schemes for Mixed Media. Proc. of the 2nd IEEE Pacific-Rim Conference on Multimedia (IEEE-PCM), Beijing, Oct 2001

[Yuan01] W. Yuan, K. Nahrstedt, X. Gu. Coordinating Energy-Aware Adaptation of Multimedia Applications and Hardware Resource. Proceedings of the 9th ACM Multimedia (Middleware Workshop), October 2001.

[Yuan02] Yuan, M., Lonf, J. Build database-powered mobile applications on the Java platform. Javaworld, http://www.javaworld.com, January 2002

[Zenal97] ZENEL, B., AND DUCHAMP, D. A General-purpose Proxy Filtering Mechanism Applied to the Mobile Environment. In Proceedings of MobiCom '97 (Budapest, Hungary, Oct. 1997).

[Zhang93] Zhang, L., Deering, S., Estrin, D., Shenker, S. RSVP: a new resource reservation protocol. IEEE Network, Vol. 7, No. 5, pp. 8-18, September 1993

[Zhao01] Zhao, Xinhua, Castelluccia, Claude and Baker, Mary. Flexible Network Support for Mobile Hosts. Mobile Networks and Applications (MONET), volume 6, number 2, March/April 2001.

[Zhao02] Zhao, B. Y., Joseph, A., And Kubiatowicz, J. Locality aware mechanisms for large-scale networks. In Proc. of Workshop on Future Directions in Distributed Computing, June 2002

[Ziv77] Ziv, J and Lempel, A. A universal algorithm for sequential data compression. IEEE transactions on information theory, IT-23,3 , pp 337-43, May 1977

Glossary of Terms

AIFF: Audio Interchange File Format

Aliasing: A term used to describe the unpleasant jaggy appearance of unfiltered angled lines. Aliasing is often caused by sampling frequencies being too low to faithfully reproduce an image. There are several types of aliasing that can affect a video image including temporal aliasing (e.g., wagon wheel spokes apparently reversing) and raster scan aliasing (e.g., flickering effects on sharp horizontal lines).

Alpha channel: Colour in an RGB video image is stored in three colour channels. An image can also contain a *matte* (also known as a *mask*) stored in a fourth channel called the "alpha channel."

Analog: The principal feature of analog representations is that they are continuous. For example, clocks with hands are analog—the hands move continuously around the clock face. As the minute hand goes around, it not only touches the numbers 1 through 12, but also the infinite number of points in between. Similarly, humans experience of the world, perceived in sight and sound, is analog. Humans perceive infinitely smooth gradations of light and shadow; infinitely smooth modulations of sound. Traditional (non-digital) video is analog.

AU: Audio Format; one of the most common audio formats on the Web

Asynchronous: Not coordinated in time; contrasted with synchronous. In a typical synchronous protocol, each successive transmission of data requires a response to the previous transmission before a new one can be initiated. An asynchronous protocol allows transmissions to occur independently of one another. In computer communications using asynchronous protocols, each piece of data usually has a start bit at the beginning and a stop bit at the end, so that the valid data can be distinguished from random noise. Most communications between computers and devices are asynchronous; the public Internet is based on an asynchronous system.

AVI: Defined by Microsoft, "AVI" stands for *Audio Video Interleave*. AVI is the standard file format for digital video on the Microsoft Windows platform. AVI is not a streaming format; ASF (*Advanced Streaming Format*) is the Windows Media streaming format.

Bandwidth: the amount of data that can be transferred from one point to another, usually between a Web server and a Web Browser; It is a measure of the range of frequencies a transmitted signal occupies. In digital systems, bandwidth is the data speed in bits per second. In analog systems, bandwidth is measured in terms of the difference between the highest-frequency signal component and the lowest-frequency signal component.

Bit: short for binary digit; the smallest unit of computer data

Bps: bits per second; common units of measurement for data transfer rates

Broadband: the telecommunication that provides multiple channels of data over a single communications medium

Broadcast quality: Standard of quality associated with current expectations of clearly-received network or cable television.

Buffer: A temporary storage area, or holding place, usually in the RAM (random access memory) of a computer or peripheral device (e.g., a printer), where data can be collected and/or held for ongoing or later processing.

Byte: Eight (8) *bits.*

CD: Reduced resolution highly compressed digital video, often with low frame rates

Channel: a separate path through which signals can flow; refers to the number of data streams that a file creates, typically one for monophonic sound and two for stereophonic sound

Codec: abbreviation for compressor/decompressor. An algorithm or scheme used when recording digital video or audio. A codec is used, for example, when video is transmitted over the Internet; the video is compressed on the sending end and decompressed on the receiving end. Users can select a codec based on the audio or image quality, and image size preferred.

Compression: the coding of data to reduce file size or the bit rate of a stream. Content that has been compressed must be decompressed for playback. A codec contains the algorithms for compressing and decompressing audio and video.

Content: data that an encoder or server streams to a client or clients. Content can originate from live audio or live video presentation, stored audio or video files, still images, or slide shows. The content must be translated from its original state into a Windows Media format before a Windows Media server can stream it. Windows Media servers can stream live streams or stored Window Media files as content.

De-interlacing: The process performed by video editing software, of removing *interlacing* from video originally intended for display on television monitors, in order to make it suitable for display on computer monitors.

Delta frame: In *interframe* (a.k.a. *temporal*) *compression*, periodic *keyframes* store all the information that comprises a frame, while delta frames store only the information that changes from frame-to-frame in between keyframes.

Encoder: A software application or a device (hardware) used to encode—i.e., compress and format (see *encoding*) digital video or audio.

Encoding: Encoding accomplishes two main objectives: 1) it reduces the size of video and audio files, by means of compression, making Internet delivery feasible, and 2) it saves files in a format that can be read and played back on the desktops of the targeted audience. Some encoding solutions may also be configured to provide additional processing functions, such as digital watermarking, for example. Encoding may be handled by a software application, or by specialised hardware with encoding software built in. Note that the term *compression* is often used interchangeably with the term "encoding" when referring to the final step in preparing media files for Web distribution; but compression is only a part of the encoding process.

8-bit/16-bit: the length of each segment of data, akin to "word length" in file

Fps: number of frames per second that a video file displays

Hinted movie: Term used in Apple's QuickTime architecture for video files that are formatted for *true streaming*.

Interlacing: System developed for early television and still in use in standard television displays. To compensate for limited persistence, the electron gun used to illuminate the phosphors coating the inside of the screen alternately draws even, then odd horizontal lines. By the time the even lines are dimming, the odd lines are illuminated. Humans perceive these "interlaced" *fields* of lines as complete pictures.

Intra-frame compression: Also known as *spatial compression*, intra-frame compression reduces the amount of video information within each frame individually.

ISDN: Integrated Serial Digital Network; a set of standards for digital transmission over ordinary telephone copper wire as well as over other media with two levels of service, i.e. Basic Rate Interface and Integrated Service Digital Network

ISO MPEG-4 video codec version 1.0: A type of codec based on the ISO MPEG-4 standard. It enables you to encode content produced by many consumer electronics devices, such as digital video cameras and cell phones.

JPEG: a graphic image created by utilizing compressed qualities; quality of image specified with suffix ".jpg"

KB: Kilobytes

Kbps: kilobits per second, or thousands of bits per second; a measure of bandwidth on a data transmission medium

Lossless compression: data compression techniques in which no data is lost.

Lossy compression: Refers to data compression techniques in which some amount of data is lost. Lossy compression technologies attempt to eliminate redundant or unnecessary information.

Mbps: millions of bits per second or megabits per second; a measure of bandwidth on a data transmission medium (i.e. twisted-pair copper cable, optical fibre, coaxial cable)

Megabit: a million bits, commonly used for measuring the amount of data that is transferred in a second between two telecommunications points

Megabyte: a million bytes; a measure of computer processor storage and real and virtual memory

M-Bone: The MBONE (Multicast Backbone) is a virtual network layered on top of the physical Internet to support the routing of multicast packets. For more information, see www.mbone.com.

MBR: Short for *multiple (or multi) bit rate*, and also known as *adaptive streaming,* MBR is a technique by which several streams, compressed at different bit rates, are encoded together into a single file. When the client calls for the media file, a negotiation between client and server determines the available bandwidth, and the appropriate stream is transmitted.

Media: A term with many different meanings, in the context of *streaming media*, it refers to video, animation, and audio. The term "media" may also refer to something used for storage or transmission, such as tapes, diskettes, CD-ROMs, DVDs, or networks such as the Internet.

Media server: Specialized server software that takes advantage of appropriate Web transfer protocols such as RTSP (real time streaming protocol), as well as special communication techniques between clients and servers, to facilitate the continuous playback of synchronized audio and video in real time, adjusting the streams transmitted to the actual bandwidth available. Media server software may be running on discrete hardware, or can be deployed in combination with Web server software running on the same device.

Megahertz (MHz): a unit used to measure a computer's speed

Modem: modulates outgoing digital signals from a computer or other digital device to analog systems for a conventional copper twisted-pair telephone line; also demodulates the incoming analog signal and converts it to a digital signal for the digital device

MP3: MPEG-1 Audio Layer-3, or a standard technology and format for compressing a sound sequence into a very small file while preserving the original level of sound quality when it is played, i.e. ".mp3."

MPEG: the Moving Picture Experts Group who develops standards for digital video and digital audio compression. (To use MPEG video you need a personal computer with sufficient processor speed, internal memory, and hard disk space, as well as an MPEG viewer or software that plays MPEG files.)

Multicasting: sending the same file to multiple users at the same time

Multiple bit rate video: the support of multiple encoded video streams within one media stream. By using multiple bit rate video in an encoder, you can create media-based content that has a variety of video streams at variable bandwidths ranging from e.g., 28.8 Kbps through 300 Kbps, as well as a separate audio stream. After receiving this multiple encoded stream, the server determines which bandwidth to stream based on the network bandwidth available. Multiple bit rate video is not supported on generic HTTP servers.

Narrowband: Low-bandwidth (typically dial-up) network connections usually 56 Kbps or lower.

Narrowcast: Transmission of media to multiple end-users but, as differentiated from *broadcast*, not to everyone on a network

NMEA GPS Interface with Other Electronics Devices: NMEA is an abbreviation for the National Marine Electronics Association, the group that establishes the data protocol and wiring standards for the marine electronics industry. Some GPS units can receive DGPS data from beacon and FM receivers. GPS receivers must also be able to send standard positioning and navigational information to a variety of listener devices such as charting instruments, autopilots, and others. Most quality built GPS products permit their users to select from two different NMEA data protocols that transmit data output sentences. The first protocol is NMEA 0180, which is reserved strictly for sending steering information, primarily to marine auto pilots. And the second protocol, NMEA 0183, sends latitude/longitude position, steering, speed, and other navigational data.

On-demand: Media which is not transmitted live, as it is recorded, but is made available as an archive on a server, for end-users to access when they wish. A television broadcast is "live;" renting a video and watching it at home is "on-demand."

Plug-in: A plug-in extends the capabilities of a Web browser, enabling the client to display or playback a file type which the browser itself cannot handle.

Pre-processing: Sometimes called *optimizing*, pre-processing removes non-essential information from video and audio—information that is difficult to encode and/or does not substantively add to the quality of the streamed media. So pre-processing prior to encoding reduces the burden on the compressor, potentially saving time and CPU capacity.

Post-production: The phase of a film or video project that involves editing and assembling footage and adding effects, graphics, titles, and sound.

Pre-production: The planning phase of a film or video project—usually completed prior to commencing production.

RTSP: Real Time Streaming Protocol Sampling Frequency: the number of times an audio file is sampled in a given period of time

Server cluster: A group of networked servers that streamline internal processes by distributing the workload and expedite computing processes by harnessing the power of multiple servers. Load balancing software tracks demand for processing power from different machines, prioritizing the tasks, and scheduling and rescheduling them depending on priority and demand users put on the network. Redundancy ensures that if one server in the farm fails, another can step in as backup.

SECAM: Similar to *PAL* at 25 FPS, the SECAM format is employed primarily in France, the Middle East, and Africa. It is only used for broadcasting. In countries employing the SECAM standard, PAL format cameras and decks are used.

Skin: A custom GUI (graphical user interface) designed for a specific media player.

Streaming Video: a sequence of moving images that are transmitted in compressed form over the Internet and displayed by a viewer as they arrive; is usually sent from pre-recorded video files, but can be distributed as part of a live broadcast feed.

Streaming Media: streaming video with sound

True streaming: Affording real-time access to content via the Internet or an intranet, true streaming is enabled by a specialized server application that relies on streaming protocols to adjust the rate of transmission to accommodate available bandwidth.

Tunneling: The use of specially designed paths to carry multicast traffic over the Internet.

24-bit colour: Type of colour representation used by current computers. For each of the Red, Green, and Blue components, 8 bits of information are stored and transmitted—24 bits in total. With these 24 bits of information, over a million different variations of colour can be represented.

VOD: a file that a user can download to watch on demand

(.wav): the Windows standard for waveform audio files

Web casting: The technique of broadcasting media over an intranet, extranet, or the Internet.

Web server streaming: Another term for HTTP streaming, pseudo-streaming, or progressive download.